W0106096

Current Topics in Microbiology
136 and Immunology

Editors

A. Clarke, Parkville/Victoria · R.W. Compans,
Birmingham/Alabama · M. Cooper, Birmingham/Alabama
H. Eisen, Paris · W. Goebel, Würzburg · H. Koprowski,
Philadelphia · F. Melchers, Basel · M. Oldstone,
La Jolla/California · P.K. Vogt, Los Angeles
H. Wagner, Ulm · I. Wilson, La Jolla/California

The Molecular Biology of Bacterial Virus Systems

Edited by G. Hobom and R. Rott

With 20 Figures

Springer-Verlag
Berlin Heidelberg New York
London Paris Tokyo

Prof. Dr. G. Hobom
Institut für Mikrobiologie und Molekularbiologie
der Universität
Frankfurter Str. 107, D-6300 Giessen

Prof. Dr. R. Rott
Institut für Virologie der Universität
Frankfurter Str. 107, D-6300 Giessen

ISBN-13:978-3-642-73117-4   e-ISBN-13:978-3-642-73115-0
DOI: 10.1007/978-3-642-73115-0

This work is subject to copyright. All rights are reserved, whether the whole
or part of the material is concerned, specifically the rights of translation, reprint-
ing, reuse of illustrations, recitation, broadcasting, reproduction on microfilms
or in other ways, and storage in data banks. Duplication of this publication
or parts thereof is only permitted under the provisions of the German Copyright
Law of September 9, 1965, in its version of June 24, 1985, and a copyright
fee must always be paid. Violations fall under the prosecution act of the German
Copyright Law.

© Springer-Verlag Berlin Heidelberg 1988
Softcover reprint of the hardcover 1st edition 1988
Library of Congress Catalog Card Number 15-12910

The use of registered names, trademarks, etc. in this publication does not imply,
even in the absence of a specific statement, that such names are exempt from
the relevant protective laws and regulations and therefore free for general use.

Product Liability: The publisher can give no guarantee for information about
drug dosage and application thereof contained on this book. In every individual
case the respective user must check its accuracy by consulting other pharmaceuti-
cal literature.

2123/3130-543210

Table of Contents

Indexed in Current Contents

List of Contributors

You will find the addresses at the beginning of the
respective contribution

Retroregulation of Bacteriophage λ *int* Gene Expression

G. GUARNEROS

1 Introduction

Upon infection of *Escherichia coli*, bacteriophage λ may elicit either a lytic or a lysogenic response. In the lytic response the infected cell is killed but produces many copies of the phage. In the lysogenic program the phage's lytic functions are repressed, and its DNA persists in the surviving cell integrated in the chromosome as a prophage. The prophage may be induced into the lytic cycle by inactivating the repression system. When this occurs, the phage DNA is excised from the host chromosome, replicated, and packaged into viral particles. Many copies of the phage are produced, as in the lytic response to infection. Both integration of phage DNA into, and excision of prophage DNA from the bacterial chromosome require the phage-directed Int protein and the bacterial integration host factor (IHF). The excision reaction also requires Xis protein encoded by the phage genome (see WEISBERG and LANDY 1983 for a review).

Timely expression of *int* and *xis* genes and the direction of the integration-excision reaction are regulated by a complex system of transcriptional, posttranscriptional, and protein activity signals (ECHOLS and GUARNEROS 1983). Two promoters, *p*I and *p*L, initiate *int* gene transcription (Fig. 1). *p*L is active immedi-

---

Department of Genetics and Molecular Biology, Centro de Investigación y de Estudios Avanzados, A.P. 14-740, Mexico City, Mexico 07000

Current Topics in Microbiology and Immunology, Vol. 136
© Springer-Verlag Berlin·Heidelberg 1988

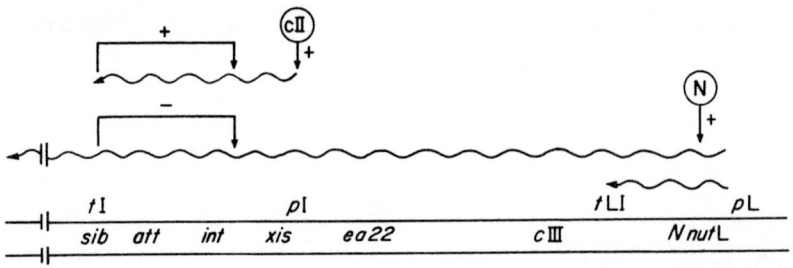

Fig. 1. Regulation of *int* gene expression. Immediately after infection, transcription initiated at *p*L terminates at *t*L1. Once active N protein is available, it acts at the site *nut*L and transcription initiated at *p*L proceeds through *t*L1 and other terminators. The *int* gene is transcribed but is not expressed from the *p*L transcript because the regulator *sib* prevents *int* mRNA translation into an active Int product. Promoter *p*I is activated by *c*II protein later in infection. The *p*I transcript terminates at *t*I and is translated efficiently into Int protein. The parallel lines represent a segment of λ DNA; several genetic markers have been positioned between the lines. *t*I overlaps *sib*, and *p*I partly overlaps *xis* in the DNA. The *wavy arrows* indicate origin, direction, and extent of λ transcripts. The *straight arrows*, flanked by + or − signs, indicate stimulatory or inhibitory activities. *Encircled* N or *c*II symbolize the respective λ-encoded proteins

ately after infection and controls several genes in the so-called λ left operon including the *xis* and *int* genes. Protein N, the product of the first gene in the λ left operon, interacts with the transcription complex through a site, *nut*L, preventing transcription termination within the left operon. Transcription from *p*L is reduced within 10 min after infection as part of the phage developmental program (FRIEDMAN and GOTTESMAN 1983). Promoter *p*I, activated by λ *c*II protein, controls *int* transcription. *p*I sequences partly overlap with the initiation codon of the preceding gene *xis* (ABRAHAM et al. 1980; HOESS et al. 1980). Therefore *c*II-stimulated *p*I-RNA lacks the translation start and other amino terminal codons for Xis protein. *c*II protein is functional only after it has been synthesized and has accumulated in the cell, which occurs sometime after infection (WULFF and ROSENBERG 1983; FIEN et al. 1984). Thus, *p*I is activated at about the time *p*L is repressed. The transcript initiated at *p*I terminates at *t*I, 276 nucleotides beyond the stop codon of the *int* gene (SCHMEISSNER et al. 1984b; OPPENHEIM et al. 1982). The *p*L transcript does not terminate at *t*I as a result of the antitermination activity of N protein (SCHMEISSNER et al. 1984a). This difference determines opposite effects on *int* gene expression for the two transcripts. The *N*-dependent antiterminated *p*L transcript is defective for Int expression whereas the *c*II-dependent *p*I transcript expresses Int efficiently. Expression or lack of expression of the *int* gene is controlled by signals downstream of *int* in the mRNA. The signal *sib* in the antiterminated *p*L transcript reduces expression of the *int* transcript. This regulatory system, initiated by an element transcriptionally distal to the target gene, has been named retroregulation (SCHINDLER and ECHOLS 1981). Another distal element to *int*, the *t*I terminator at the end of the *c*II-stimulated transcript from *p*I, provides for efficient Int synthesis. To distinguish between these two regulatory effects we propose the names "negative retroregulation" for *sib* inhibition and "positive retroregulation" for *t*I stimulation of *int* expression.

Herein is collected all the available information on gene *int* negative retroregulation and other possible instances of RNase III participation in λ gene regulation. Few other cases of gene regulation in phages other than λ have been included. I will refer also to λ *int* positive retroregulation and the possible participation of RNase III in this control. Reviews addressing the subject of retroregulation have been published previously (GOTTESMAN et al. 1982; COURT et al. 1983 b, c; ECHOLS and GUARNEROS 1983; GUARNEROS and GALINDO 1984).

2 Inhibition of *int* Gene Expression by the *sib* Region

2.1 The *sib* Region

The evidence that the *int* gene is subject to retroregulation derived initially from in vivo experiments. Phage λ cII⁻ mutants are defective in the expression of Int (KATZIR et al. 1976; CHUNG and ECHOLS 1977; COURT et al. 1977). A deletion in the *b* region of the λ genome, *b*2, suppresses this defect of cII⁻ mutants. These results indicate the presence of an inhibitor in the λ *b* region (GUARNEROS and GALINDO 1979). The inhibitor is selective for the *p*L transcript because it does not affect Int systhesis directed from the *p*I promoter. Previous but complex results had indicated effects of *b*2 on *int* expression (LEHMAN 1974; ROEHRDANZ and DOVE 1977).

In the *b* region-DNA the position of the inhibitor *sib* has been precisely determined by deletional analysis and location of *sib*⁻ point mutations. Functional analyses of deletion mutants indicate that the 251 base pairs (bp) to the left of the center of the attachment site are important for negative regulation of *int* (EPP et al. 1981; GUARNEROS et al. 1982). Additional deletions generated by *Bal*31 resection defined the left end of *sib* at position −196 (negative numbers refer to base positions to the left of the central base pair in the attachment site that arbitrarily is assigned position zero) (OPPENHEIM et al. 1982; COURT et al. 1983a). These results agree with observations made in another lambdoid phage. Phage 434 shares DNA homology with λ in a segment spanning 110 bp between positions −197 and −87 (MASCARENHAS et al. 1981). A segment of phage 434 DNA containing homology to the *sib* region causes retroinhibition of *int* (MASCARENHAS et al. 1983).

Sequencing of four *sib*⁻ point mutations reveals that all the mutations fall in a 50-bp segment of DNA with a hyphenated dyad symmetry that includes bp at positions −195 and −146 (Fig. 2). As will be discussed below, this is the size and location of a minimum estimate for the site *sib*. It certainly agrees with the left limit at position −196 as defined by deletion. The right limit of the *sib* sequence, however, remains to be determined.

The isolation and characterization of *sib*⁻ mutants has been very useful in understanding the mechanism of retroregulation. The *sib*⁻ mutants were isolated as phages able to promote integrative recombination in the absence of cII function (GUARNEROS et al. 1982; MONTAÑEZ et al. 1986). Genetic mapping of three independent *sib*⁻ isolates indicates that the mutations lie in the

4   G. Guarneros

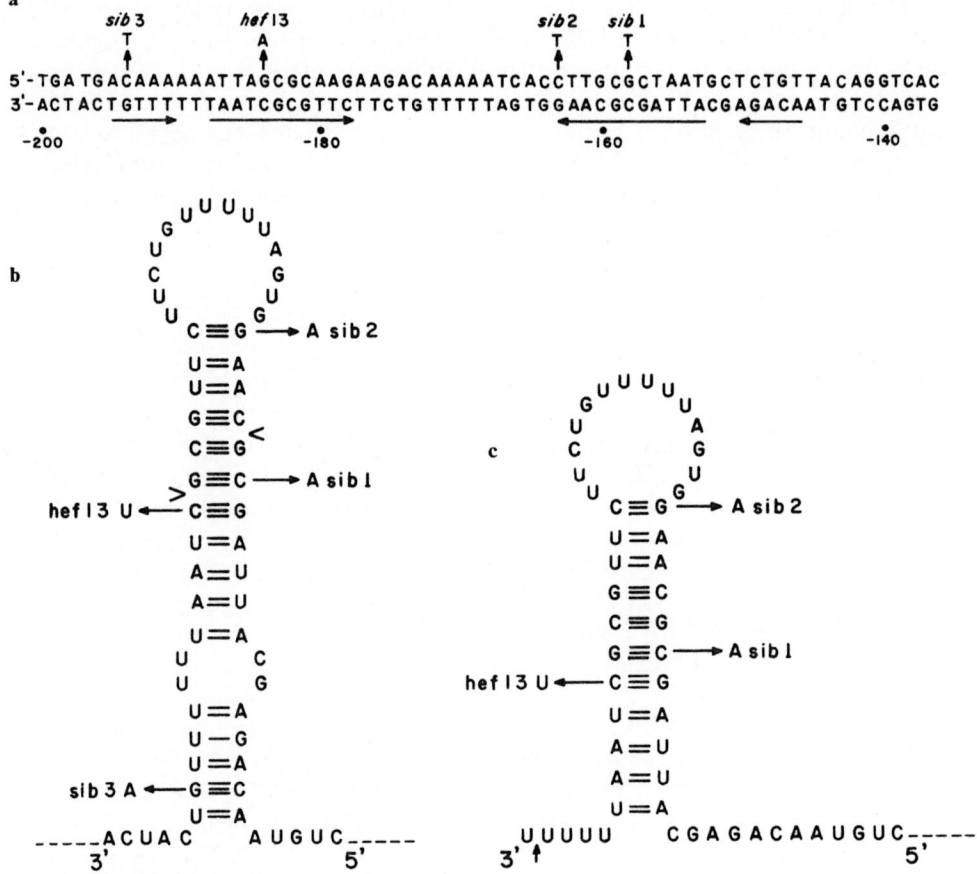

Fig. 2a–c. DNA sequence and potential RNA structures of the *sib/tI* region of λ. **a** A segment of λ DNA between positions −200 and −140 to the left of *att* is shown. The location and base changes of the three *sib* and one *hef* mutations are indicated (GUARNEROS et al. 1982). The *horizontal arrows* underline the region of dyad symmetry in *sib*. **b, c** Possible secondary structures for leftward mRNA from this region. The double stem-and-loop (**b**) can be formed by *pL* RNA but not by *pI* RNA. *Arrowheads* indicate the sites on the RNA where RNase III cuts. The *pI* transcript terminates at positions −192 or −193 in a stem-and-loop structure, followed by a run of uridylate residues (**c**) (SCHMEISSNER et al. 1984b). Note that structure **c** shares the upper stem-and-loop of structure **b**. The corresponding mutational changes are indicated in the leftward mRNA structures

241-bp DNA segment to the left of *att* in the marker order: *sib3-sib2-sib1* (MONTAÑEZ et al. 1986). Another mutant, *hef*13, was isolated independently and mapped to the left of *att* (ROEHRDANZ and DOVE 1977). The DNA sequence in the 251-bp region to the left of *att* was determined in the four mutants (GUARNEROS et al. 1982). Each has a single base substitution (Fig. 2). The changes confirmed the genetic map inferred from recombination experiments (MONTAÑEZ et al. 1986).

2.2 RNase III Processing of the *sib* Transcript

The inhibitor *sib* is a site requiring contiguousness to *int* in the same chromosome to exert its action; this was first inferred from *cis-trans* experiments by coinfection with phages *sib⁺* and *sib⁻* (GUARNEROS and GALINDO 1979; COURT et al. 1983 b; MONTAÑEZ et al. 1986). This conclusion agrees with the analysis of the sequence in the *sib* region, which, as defined by deletions and point mutations, does not encode a protein (DANIELS et al. 1983).

The host endonuclease RNase III is involved in negative retroregulation of λ integrase, the *int* gene product, by *sib*. Bacterial *rnc⁻* mutants, defective in RNase III, lack *sib* inhibition of Int synthesis (BELFORT 1980; SCHINDLER and ECHOLS 1981) or of Int activity (GUARNEROS and GALINDO 1979; EPP et al. 1981; GUARNEROS et al. 1982).

The *p*L antiterminated transcript is defective in the expression of *int* in spite of the fact that efficient expression of other genes located both upstream and downstream of *int* is observed (EPP et al. 1981; HENDRIX 1971). What causes this singular behavior? Oligonucleotide analysis of the in vivo N-dependent *p*L transcript shows that the message extends through the *t*I terminator and that it is processed at the site *sib*. Processing does not occur in *rnc⁻* bacteria (ROSENBERG and SCHMEISSNER 1982; SCHMEISSNER et al. 1984a). The *p*L transcript derived from *sib* mutants extends through *t*I but is deficiently processed in *rnc⁺* bacteria (MONTAÑEZ et al. 1986). *sib⁻* mutants are defective in negative retroregulation; thus, the absence of transcript processing correlates with *int* gene expression from *p*L. Transcripts made in vitro containing the *sib* region are processed by RNase III. Two cleavages occur 24 nucleotides apart in the *sib* RNA segment. However, in the secondary structure formed by the RNA, these cleavages are in opposite strands staggered by 2 bp (Fig. 2) (SCHMEISSNER et al. 1984a). Similar 2-bp staggered cuts have been found in T7 mRNA (the 1.1–1.3 cleavage site) and in the 30*S* ribosomal RNA precursor of *E. coli* (ROBERTSON 1982).

Infection experiments with λ *c*II⁻ phages show that the rate of *int* mRNA synthesis from *p*L is independent of *sib*. However *sib⁺* reduced the chemical stability of the *int* transcript, presumably as a result of RNase III processing (GUARNEROS et al. 1982).

2.3 Exonucleolytic Degradation of the Processed *int* Transcript

Some inferences about the nature of post-transcriptional inhibition by *sib* derive from the analysis of Int-amber fragments in infection experiments. Synthesis of complete Int protein is severely reduced by *sib*, but production of aminoterminal Int fragments escapes *sib* regulation (SCHINDLER and ECHOLS 1981). Consistent with these results are those of *int* mRNA 3′-end analysis by S1 mapping. In λ *c*II⁻-infected cells, the predominant 3′-end of the *sib⁺* mRNA is located within the *int* gene sequence. In contrast, infections of RNase III-defective cells with *sib⁺* phage, or of wild-type cells with *sib⁻* phage result in full length *int* transcripts (PLUNKETT and ECHOLS, personal communication). These data suggest that RNase III cleavage at *sib* allows degradation of mRNA, probably by an exonuclease activity, to a pausing or stopping site within the *int* transcript. There is evidence for the participation of two host RNA 3'-exonucleases in negative retroregulation of *int*. RNase II and polynucleotide phosphorylase (PNPase) degrade RNA progressively in the 3' to 5' direction. RNase II frees 5'-nucleoside monophosphates by hydrolysis, while PNPase yields 5'-nucleoside diphosphates by phosphorolysis of RNA. It should also be noted that RNase II degradation is blocked by RNA secondary structures (GUPTA et al. 1977; MOTT et al. 1985; PLATT 1986; BELASCO et al. 1985). This feature may explain the pause in *sib*-dependent RNA degradation within the *int* sequence (PLUNKETT and ECHOLS, see above). Bacterial mutants *rnb*⁻, defective in RNase II (SCHINDLER and ECHOLS, personal communication), and *pnp*⁻, lacking PNPase (L. OROZCO and G. GUARNEROS, unpublished results), are both deficient for *sib* inhibition of *int*. Other examples of 3' to 5' exonucleolytic degradation of mRNA have been recognized in *E. coli* (VON GABAIN et al. 1983; HAUTALA et al. 1979; KASUNIC and KUSHNER 1980).

Various models for *sib* inhibition of Int synthesis have been proposed in earlier papers (GUARNEROS and GALINDO 1979; BELFORT 1980; EPP et al. 1981; SCHINDLER and ECHOLS 1981; GUARNEROS et al. 1982; SCHMEISSNER et al. 1984a). Based on the evidence discussed above, the following statements can be made. Fifty nucleotides in the *sib* region of the *p*L transcript adopt a stem-and-loop secondary structure with a stem of complementary bases interrupted by a bubble of two mismatched pairs (Fig. 2). RNase III makes a 2-bp staggered cut in the upper stem of the structure and frees a 3'-end allowing for exonucleolytic degradation by PNPase and/or RNase II (Fig. 3). These enzymes degrade the transcript progressively in the 3' to 5' direction into the *int* coding sequence, inactivating its potential for full Int protein translation. Eventually the exonuc-

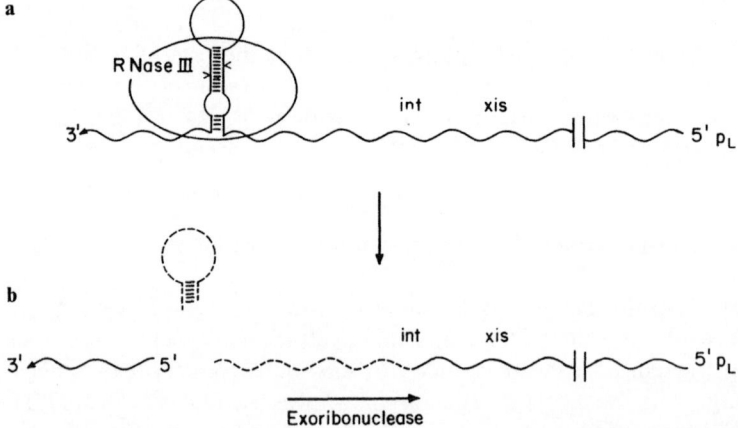

Fig. 3a, b. Model for negative retroregulation. **a** Endonucleolytic cleavage by RNase III on the *sib* secondary structure of the antiterminated *p*L transcript. **b** 3' to 5' exoribonuclease degradation of the cleaved transcript to inactivate the *int* transcript. This degradation stops, or pauses, within *int* without affecting *xis*. See text for details. The *wavy arrow* indicates the *p*L-initiated RNA. The *ellipse* represents RNase III. The *discontinuous lines* in **b** indicate RNA degradation leolytic degradation stops or pauses at specific sites in the mRNA. The processing of *sib* RNA by RNase III indicates that, indeed, the *sib* transcript adopts a hairpin-structure, since RNase III requires about 20 bp of a double-stranded RNA helix to recognize and cut stem-and-loop structures (ROBERTSON 1982).

2.4 *sib* Inhibition Is Limited by Distance

The presence of the *sib* region causes a severe inhibition of Int synthesis from the antiterminated *p*L transcript; however, there is little or no retroregulation on the synthesis of the proteins from the nearby upstream *xis* and *ea*22 genes (SCHINDLER and ECHOLS 1981). The *ea*22 gene is retroregulated when the DNA between *sib* and *ea*22 is deleted. Similarly if *int* is moved farther away from *sib* by insertion of a segment of bacterial DNA as short as 1000–2000 bp, retroregulation is eliminated (GUARNEROS et al., unpublished data). Based on these results it is suggested that the proximity from the processed site to the target gene is critical for negative retroregulation. As proposed earlier, secondary structures in the mRNA may be responsible for the distance effect by blocking exonuclease progression. Whether the expression of a distant gene occurs before the proposed 3′ to 5′ degradation inactivates its transcript or the exonuclease encounters a barrier that stops or delays degradation remains to be determined.

Other bacterial genes can also be inhibited by *sib*. A 242-bp DNA fragment containing *sib* cloned beyond the *gal*K gene in an appropriate plasmid vehicle and host cell causes inhibition of *gal*K expression (SCHMEISSNER et al. 1984a). In a similar construction, a 494-bp fragment carrying *sib*, *att*, and the end sequences of *int*, cloned downstream to an *int-lacZ* fusion, prevents expression of *β*-galactosidase (L. KAMEYAMA and G. GUARNEROS, unpublished data). These constructions allow the measurement of *sib* inhibition by enzymatic assays.

3 Positive Retroregulation

In the previous sections, evidence was reviewed that shows inhibition of Int expression from the antiterminated *p*L transcript by *sib*. Int expression from the *p*I message is also regulated. In this case the *t*I terminator that overlaps *sib* in λ DNA unexpectedly stimulates Int expression. This activation will be called positive retroregulation.

In vivo experiments have shown that Int from *p*I is more efficiently expressed in the presence of a wild-type *sib/t*I region than in mutants, as measured by integrative recombination (MILLER et al. 1981; COURT et al. 1983b; GUARNEROS and GALINDO 1984) or by Int antigens (BEAR and COURT, personal communication). In vivo *p*I transcripts terminate efficiently at *t*I (OPPENHEIM et al. 1982; SCHMEISSNER et al. 1984b). Small cloned segments containing *t*I cause transcription to terminate in vectors designed specifically to monitor termination (LUK et al. 1982; SCHMEISSNER et al. 1984b). In vitro transcription also terminates at the *t*I site at positions −192 and −193 from *att* (Fig. 2) (ROSENBERG and

SCHMEISSNER 1982; SCHMEISSNER et al. 1984b). The DNA sequence at *t*I corresponds to a segment containing 50% of G–C bp with dyad symmetry followed by a stretch of thymidylate residues at the 3′-end. Such sites are characteristic of rho-independent transcription terminators in bacteria (ROSENBERG and COURT 1979; PLATT 1986).

There are differences between the *p*I and *p*L transcripts in the *sib* region that may bear on how *int* is expressed. Since *sib* and *t*I sites overlap at the same symmetric segment of DNA, antiterminated *p*L and terminated *p*I transcripts could form stem-and-loop structures which are nearly identical. The *t*I-terminated transcript, however, lacks 3′-sequences, which prevent it from forming the lower stem of the *sib* structure (Fig. 2). This fact may explain the high susceptibility of *sib* RNA and the resistance of *t*I RNA to cleavage by RNase III (ROSENBERG and SCHMEISSNER 1982; SCHMEISSNER et al. 1984b); that in turn could explain the difference of the two transcripts in negative retroregulation. sib^- point mutations or deletions that eliminate *sib* are equally efficient in blocking negative retroregulation of *p*L transcripts (MONTAÑEZ et al. 1986). In contrast, these same mutations affect *int* expression from *p*I transcripts differentially (Table 1) (COURT et al. 1983b). Int activity in λ^+ and $\lambda sib3$ infections is high, $\lambda sib2$ infections yield intermediate levels, whereas $\lambda sib1$ and the deletion *b*2 produce the lowest Int activity. The lack of effect of $\lambda sib3$ infection was expected because the mutation lies outside the *p*I-terminated transcript, so it should behave like λ^+. The defect of *sib*1, *sib*2, and *b*2 mutations in inducing full Int expression shows that the integrity of *t*I is essential in positive retroregulation. However, there is no strict correlation between positive retroregulation and in vivo termination. The efficiency of transcription termination

Table 1. Effect of *sib* mutations and RNase III on positive retroregulation of *int*

sib allele in infecting phage[b]	Int activity[a]	
	rnc^+	rnc^-
sib^+ or *sib*3	very high	high
*sib*2	high	low
*sib*1 or *sib*Δ	low	very low

[a] Int activity was measured by an in vivo assay based on Int-dependent production of λ phage by a lysogen following superinfection by a heteroimmune phage (GUARNEROS and GALINDO 1979). Two consecutive levels of activity differ about ten fold: very high, high, low, very low. The *E. coli* strains were *rnc*105 (defective in RNase III) and rnc^+ (the isogenic wild type) (from A. OPPENHEIM and M. GOTTESMAN).

[b] The infecting phages carried one of the indicated *sib* alleles. *sib*Δ was the *b*2 deletion. The active promoter for *int* expression was *p*I. This was achieved by using *nut*L mutants, unable to express *int* from *p*L promoter.

for the *sib⁻* mutations was assayed in vivo in a plasmid system (MCKENNEY et al. 1981). All three *sib⁻* mutations terminated nearly as efficiently (95%–99%) as the *sib⁺* allele (MONTAÑEZ et al. 1986). Thus positive retroregulation is more sensitive than transcription termination to mutations in the 3′ stem-and-loop structure.

3.1 The Role of RNase III

The host RNase III participates in negative retroregulation of Int expression from the antiterminated *p*L transcript (see above). Does RNase III play a role in the expression of Int from *p*I? Integration experiments show that Int activity from *p*I in bacteria deficient in RNase III was reduced in relation to the levels observed in the wild-type host. This effect was seen for both *sib⁺* and *sib⁻* cases (compare columns 2 and 3 in Table 1). We propose that RNase III stimulates Int activity by stabilizing terminated *p*I transcripts. RNase III protein may bind to the terminator structure at the 3′-end of the message, preventing its exonucleolytic degradation (Fig. 4a). *t*I structures disrupted by mutational mismatches (*sib*1 and *sib*2) are poorly protected by RNase III protein (Fig. 4b). The remaining Int activity noticed in the *rnc⁻* host may reflect residual protection by the mutated RNase III. This possibility is supported by the recent

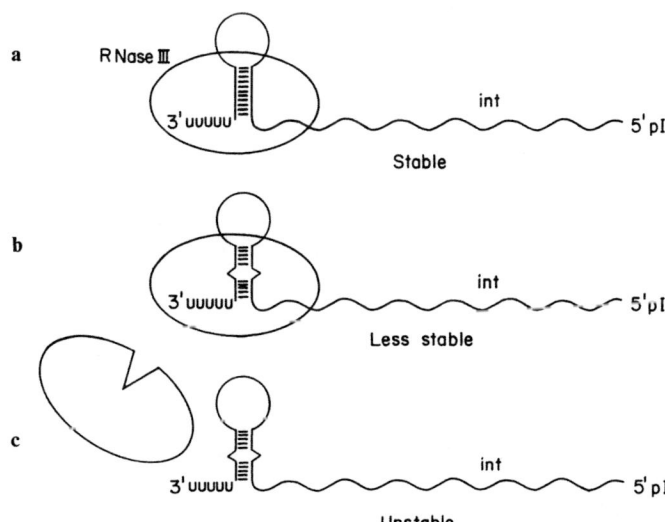

Fig. 4a–c. Model for positive retroregulation and effect of *sib* mutations. **a** RNase III interacts with the potential stem-and-loop structure at the 3′-end of the terminated *p*I transcript but does not process. In fact this complex protects *int* mRNA from 3′ nucleolytic degradation. **b** A mutation in the stem-and-loop of the terminated transcript that introduces a mismatch in the stem reduces the affinity of the structure to RNase III and the stability of the transcript. **c** A combination of mutations in the transcript and in the RNase III results in an unstable *int* transcript. *Symbols as in Fig. 3*

demonstration that the mutation used in these experiments is a 1-bp change encoding a partly active RNase III (PANAYOTATOS and TRUONG 1985; NASHI-MOTO and UCHIDA 1985). Other proposals involving an indirect role of RNase III in the stimulation of Int activity are not ruled out.

4   Developmental Control of *int* by Retroregulation

How does retroregulation account for *int* expression at different phases of λ development? Little Int is synthesized early after infection either from pL or pI. pL initiates transcription of *int*, but *sib* inhibits expression of the *int* transcript. Int synthesis from pI cannot start because cII protein, essential to initiate pI transcription, has not yet been made. Int function, however, is not needed early after infection, when the initial events of λ development are common for both lytic and lysogenic responses. Thus, the potential disadvantage of early Int expression is avoided (COURT et al. 1983b, c). Apparently this negative control on *int* is not essential for lytic development. Unexpectedly, λ *sib*⁻ mutants, that express Int early after infection, show a two- to fivefold increase in phage yield upon infection in relation to *sib*⁺ phages (GUARNEROS, MONTAÑEZ, CAMARENA, and HERRERA, unpublished results). Lysogenization by λ requires two phage-directed proteins: cI repressor and Int. During λ development, entering either the lysogenic or the lytic pathways depends on the levels of cII protein at the critical moment (WULFF and ROSENBERG 1983). High levels of cII protein activate transcription initiation of both cI and *int*, ensuring the coordinate synthesis of both proteins. In addition, Int expression from the terminated pI transcript is very efficient (see above). Therefore, ample Int is available to ensure integration and lysogeny. Positive retroregulation does affect lysogeny, if modestly, since we find that *sib*⁺ phages lysogenize twice as frequently as *sib*⁻ phages after infection (GUARNEROS and HERNANDEZ, unpublished results).

One could predict that early synthesis of Int in a λ *sib*⁻ infection would be harmful to the cell by causing early integration of λ DNA into the host chromosome and cell killing by λ-driven DNA replication (COURT et al. 1983b). This apparently is not the case, as cell-killing rates following λ*sib*⁺ or λ*sib*⁻ infections are indistinguishable (GUARNEROS and HERNANDEZ, unpublished results). In the absence of *sib*, Xis, also synthesized from the pL transcript, is likely to inhibit Int function and integration (NASH 1975a, b; ABREMSKI and GOTTESMAN 1982).

Integration of the λ DNA itself affects retroregulation of the *int* gene. The site at which λ integrative recombination occurs, *att*, is located between *sib* and *int* in the vegetative map of the phage (Fig. 1). When λ DNA integrates in the host chromosome, *sib* and *int* are dissociated to opposite ends of the prophage (Fig. 5). After induction, prophage excision requires Int and Xis proteins as an early step in lytic development. Whereas synthesis of Int from pL promoter is inhibited by *sib* immediately after infection, the absence of *sib* next to *int* in the prophage allows early Int synthesis upon induction (OPPENHEIM et al. 1982).

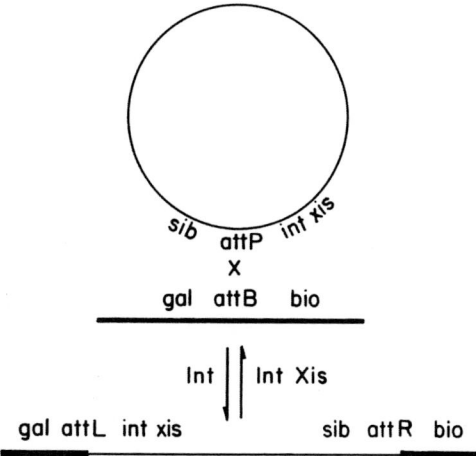

Fig. 5. Dissociation of *sib* and *int* by integrative recombination. Integrative recombination between the attachment site in the circular λ DNA (*att*P) and the attachment site in the bacterial DNA (*att*B) between the *gal* and *bio* operons results in prophage insertion and dissociation of *sib* from *int* to opposite ends of the prophage. Excisive recombination between the left and right attachment sites (*att*L and *att*R) after prophage induction reverses the integrative reaction, reestablishing *sib* next to *int*. Both reactions require Int protein and the integration host factor, IHF. Excisive recombination requires, in addition, the phage-encoded Xis protein

5 Other RNase III Processing Sites in λ Transcripts and Their Possible Regulatory Role

Transcripts initiated at *p*L are processed by RNase III. This was first inferred because λ infections of wild-type bacteria result in *p*L transcripts lacking the *N* gene sequence; in the RNase III-deficient host, however, the *p*L transcripts do contain the *N* sequence at its 5'-proximal end (LOZERON et al. 1976).

In vivo *p*L transcripts, either terminated in the absence of an active N protein (ANEVSKI and LOZERON 1981) or antiterminated by N (WILDER and LOZERON 1979), are processed by RNase III to yield two RNA fragments. One contains the 5'-triphosphate terminus, called l_1^o, and the second contains the adjacent (l_1) RNA (Fig. 6). A third RNA segment of the transcript terminated at the *t*L1 site contains the N message, which is rapidly degraded (ANEVSKI and LOZERON 1981). The antiterminated mRNA undergoes cleavage near the *t*L1 termination signals beyond gene *N* and also generates a rapidly decaying N mRNA. In the absence of RNAse III, this latter N-mRNA segment is more stable (WILDER and LOZERON 1979). A possible RNase III processing site has been identified between *ral* and *Ea*10 by S1 mapping of in vivo *p*L transcripts (HYMAN and HONIGMAN 1986). In this study no evidence for an RNase III cut near the *t*L1 signal was presented. Since HYMAN and HONIGMAN's experiments detect only stable messages, these results do not necessarily disagree with the processing site proposed by WILDER and LOZERON (1979). The possibility of retroregulation

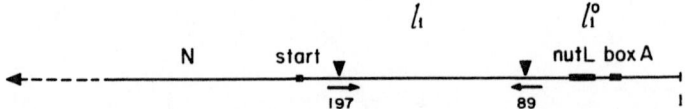

Fig. 6. RNase III processing sites in the *p*L proximal transcript. The *line* represents the initial few hundred nucleotides of the *p*L transcript. The *arrow* at the left-hand end indicates the direction of transcription. The start point of the transcript is marked *1* at the right-hand end. Other *numbers* indicate the nucleotide positions of adjacent RNase III processing sites (LOZERON et al. 1983). The *thick line segments* represent position and relative length of the respective signals (FRANKLIN 1985)

of *N* gene expression by 3′-exonuclease degradation of the *N* transcript from a downstream RNase III cut has been suggested (LOZERON et al. 1983).

The sequence of the leader RNA at the 5′-end of the *p*L transcript carries several structural and functional signals. Box A and *nut*L are sequences essential for N antitermination. In addition, two inverted repeats have been proposed as targets for RNase III (Fig. 6) (FRANKLIN 1985). Presumably, processing at these sites yields l_1^0 and l_1 RNA segments (LOZERON et al. 1976, 1983). It has been proposed that RNase III processing at these sites enhances *N* gene expression by unmasking RNA sequences essential for efficient translation (LOZERON et al. 1983). To test this supposition, gene fusions of *λ N* and *E. coli lacZ* genes have been constructed. In a host carrying an active *rnc* gene, these fusions express *β*-galactosidase activity using *N* gene translation signals. In the absence of an active *rnc* gene, however, little *β*-galactosidase activity is detected. This result is interpreted to mean that RNase III processing of the *p*L leader sequence is necessary for the efficient translation of the *N* gene (KAMEYAMA and GUARNEROS, unpublished results). *λ* phage development proceeds, albeit poorly, in an RNase III⁻ host (ANEVSKI and LOZERON 1981), and thus N function must be synthesized in this host. Whether this phenotype reveals leakiness of the *rnc*⁻ mutation or partial N expression from the unprocessed *p*L transcript is not clear yet.

Genetic and biochemical studies indicate the participation of RNase III in the initiation of *λ* gene *c*III translation. A nucleotide sequence upstream (5′) to the ribosome-binding site and the initiation codon of *c*III contains an RNase III cleavage site. A gene fusion between *c*III and *E. coli lacZ* gene does not express *β*-galactosidase activity in RNase III-deficient hosts (ALTUVIA et al. 1987). As in the *λ* gene *N* case, this result suggests that processing upstream of a gene transcript helps translation initiation, probably by removing RNA secondary structures.

The major rightward "late" RNA of bacteriophage *λ* is initiated at the *p*R′ promoter. In the absence of Q protein, the late antiterminator protein, a short (6*S*) terminated transcript is synthesized. Q antiterminates the 6*S* transcript. This Q-dependent mRNA is processed near its 5′-end by RNase III. S1 nuclease mapping and sequencing studies have defined the 5′-terminal ends of two RNase III cleavage sites. The cuts are located about 15 and 75 nucleotides downstream from the 3′-end of the 6*S*-terminated transcript (LOZERON et al. 1983). As yet there is no evidence for the regulatory role of these processing events in the expression of *λ* late proteins.

The participation of RNase III in the expression of other λ genes is indicated by the distinct protein patterns of phage-induced proteins in wild type and RNase III deficient hosts hosts, as cited in LOZERON et al. (1983).

6 Retroregulation in Other Systems

T7 gene *1.2* is negatively retroregulated. The early region of T7 DNA is transcribed by *E. coli* RNA polymerase. The primary transcript is processed by RNase III at five intercistronic regions (DUNN 1976). RNase III cleavage of the early T7 mRNA at two intergenic hairpin structures yields an intervening RNA encoding two proteins, 1.1 and 1.2 (STUDIER et al. 1979; SAITO et al. 1980). The distal recognition site is cleaved twice, 29 nucleotides apart, by RNase III. Here, as in the λ*sib* site, the cuts lie opposite each other in the postulated hairpin; the *a* site on the 5′ side and the *b* site on the 3′ side (ROBERTSON et al. 1977) (Fig. 7). In vivo the *b* site is almost always cleaved, while the *a*

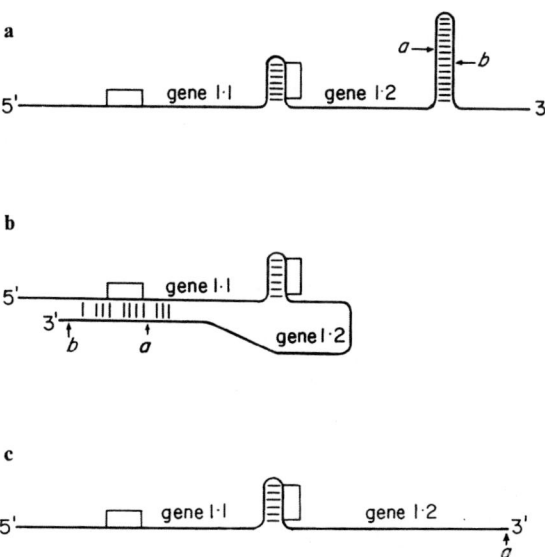

Fig. 7a–c. A model for the regulation of gene 1.2 (SAITO and RICHARDSON 1981). **a** Early T7 RNA unprocessed at the distal RNase III recognition site can form a stable hairpin structure. The ribosome-binding site for the gene *1.1* remains open, and gene *1.1* is translated. Ribosomes traversing the gene *1.1* transcript may dissociate the hairpin-associated, ribosome-binding site of gene *1.2*, allowing coordinated translation of genes *1.1* and *1.2*. **b** Cutting of the *b* site at the RNase III recognition site allows the 3′-end of the mRNA to hybridize to the complementary sequence around the ribosome-binding site of gene *1.1*, preventing its translation. The translation of gene *1.2*, dependent on gene *1.1* translation, is prevented too. **c** Cleavage of both *a* and *b* sites generates a short mRNA that lacks the hybridizing sequences to the *1.1* ribosome-binding site. Thus, translation occurs efficiently. The *boxes* represent the ribosome-binding sites. The ribosome-binding site for gene *1.2* is marked in a short hairpin structure; the long hairpin represents the distal RNase III recognition site. The relative positions of the cutting sites *a* and *b* are marked

site is cleaved about 40% of the time (DUNN 1976; ROBERTSON et al. 1977). As a consequence, the mature *1.1* mRNA can assume two forms, a larger one arising from cuts at the *b* site and a smaller one arising from cuts at both the *a* and *b* sites. Small form *1.1* mRNA is active in the expression of gene *1.2*; large form *1.1* mRNA is inactive. In the absence of processing, either as a result of mutations that affect the cleavage site or as a consequence of a mutation in the structural gene for RNase III, the gene *1.2* is expressed. The RNase III recognition site is not necessary for the expression of gene *1.2*. A mechanism of retroregulation of the *1.1* and *1.2* genes has been proposed (SAITO and RICHARDSON 1981) (Fig. 7). The sequence at the ribosome-binding site of the *1.1* gene is complementary with the nucleotides at the 3′-end of the long transcript. This transcript may fold back and form a secondary structure at the start of the *1.1* gene coding sequence, blocking its translation (Fig. 7b). Two different situations lead to protein synthesis. When wild-type T7 infects RNase III⁺ hosts, the transcript is processed at sites *a* and *b* to yield the small form of the *1.1* transcript which lacks the inhibitory end (Fig. 7c). In an RNase III-deficient host no processing occurs; the complementary sequence remains as part of the RNase III substrate structure and thus is not available for base pairing with the *1.1* coding region. Failure to translate *1.1* prevents translation of gene *1.2*. Models to explain this polarity have been set forth previously (SAITO and RICHARDSON 1981).

The expression of the T7 gene *0.3*, the first gene in the early mRNA just as *N* is the first gene in the λ *pL* transcript, requires RNase III processing for efficient protein synthesis. This result has been found both in vivo and in a cell-free protein synthesizing system (DUNN and STUDIER 1975). The mechanism for this activation is not understood. It has been suggested that the RNase III cleavage could increase the efficiency of translation initiation by exposing the ribosome-binding site in the message (STEITZ and BRYAN 1977).

The *ant* gene of phage P22 is expressed from the *pANT* transcript but not from the antiterminated *pLATE* transcript (SUSSKIND and YOUDERIAN 1983). Mutations in *tANT*, a terminator for the *pANT* transcript, reduce the rate of antirepressor synthesis. Apparently an unaltered stem sequence in the terminator hairpin allows full expression of the *ant* gene (SUSSKIND and YOUDERIAN 1983). These observations are reminiscent of *int* retroregulation; whether or not the mechanisms are similar in both cases remains to be tested.

The stability of two species of φX174 polycistronic mRNA in vivo can be altered by mutating sequences beyond the genes but immediately upstream of a transcription termination site. The wild-type phage DNA contains a *cis*-acting, mRNA-stabilizing sequence; an insertion in this sequence reduces the stability of the mRNA. The half-lives of cloned φX174 genes *B* or *D* mRNAs are increased by the wild-type sequence up to fourfold in relation to the half-lives in its absence. The stabilized mRNAs are also functional as evidenced by an increased synthesis of their products. It is suggested that terminators may contain the stabilizing sequence for mRNA (HAYASHI and HAYASHI 1985).

Work on gene expression in bacterial systems other than *E. coli* supports the role of 3′ secondary structure on mRNA stability and upstream gene expression. The *cry* transcripts, encoding the insecticidal, parasporal, crystal protein of *Bacillus thuringiensis*, terminate 4 bp downstream from an inverted repeat, 137 nucleotides beyond the terminal gene codon (WONG and CHANG 1986). A DNA segment carrying this region cloned downstream (3′) to heterologous genes causes transcription termination at the *cry* terminator and gene product overproduction. Stimulation of gene expression results from mRNA stabilization. The half-lives of the gene transcripts are increased by a factor of two to three by attaching the *cry* terminator segment to the 3′-end of cloned genes (WONG and CHANG 1986). Thus the *cry* terminator-containing fragment has a positive retroregulatory effect on heterologous genes.

Sequences distal to the *gly*A gene in *E. coli* affect the expression of the gene. A Mu phage inserted between the end of the *gly*A structural gene and its proposed transcription termination site reduces to 30% the normal level of the gene product activity (serine hydroxymethyl transferase) (PLAMANN and STAUFFER 1985). The mutation is *cis* acting and indicates that sequences distal to the *gly*A gene have a positive effect on its expression. It is not clear whether this effect is due to the long region of dyad symmetry or to the transcription termination structure (or both) that follow the translation stop codon of the gene.

Positive retroregulation has been reported in yeast. The *CYC1* gene of *Saccharomyces cerevisiae* encodes for iso-1-cytochrome *c*. A 38-bp deletion affecting the 3′ noncoding region of the gene leads to a reduction in the level of the protein. This deletion affects the *CYC1* transcription terminator, causing *CYC1* mRNAs to be much longer and more unstable than normal. Revertants with increased levels of iso-1-cytochrome *c* are chromosomal rearrangements, with a new 3′ noncoding region and new 3′ mRNA termini. These rearrangements increase the stability of *CYC1* mRNA and change mRNA translational efficiency (ZARET and SHERMAN 1984).

7 Concluding Remarks

The *sib* region of the λ genome has three different functions: inhibition of *int* expression from the *p*L transcript, stimulation of *int* expression from the *p*I transcript, and transcription termination of the *p*I transcript. The evidence indicates that the functions are independent. Likewise, overlapping functions have been recognized at a terminator in the J–F intercistronic region of bacteriophage φX174.

From the results in phages λ, P22, and φX174 and in *E. coli* and *B. thuringiensis*, corresponding mechanisms for positive retroregulation exist. These involve the participation of transcription terminators and the stabilization of the chemical and functional half-lives for the mRNAs of the controlled genes.

Our results indicate that the intact hairpin structure at the *t*I terminator in the *p*I transcript increases the activity of Int in vivo. The mechanism by which RNase III stimulates expression from the terminated *p*I transcript needs direct biochemical confirmation. Whether similar mechanisms exist in the other positively regulated sites also requires study.

Different models have been proposed for the two known cases of negative retroregulation. They both invoke processing of RNA hairpin structures by the host RNase III. The processing may be a double cut – as in the λ *sib* transcript – or a single cut – as in the distal hairpin in T7 *1.1–1.2* gene transcript. In the λ *int* case directional degradation after the initial RNase III cut eventually inactivates the *int* transcript. For T7 genes *1.1* and *1.2* folding back of the processed 3′-end of the transcript and hybridization to the ribosome-binding site of gene *1.1* may prevent the coordinated translation of the controlled genes.

Some questions related to negative retroregulation of the λ *int* gene await experimental answers. The proposed degradation of the processed *int* mRNA by 3′ exonucleases needs direct biochemical confirmation, as does the nature of the barrier that prevents *sib* inhibition from affecting upstream genes. The development of a system to search for *sib*-like inhibitors may facilitate the identification of such sites and the genes negatively retroregulated by them.

In addition to the participation of RNase III in retroregulation, a different regulatory role can be recognized for this enzyme. The cutting of 5′ leader sequences of different mRNAs correlates with efficient gene expression. This is taken to mean that RNase III processing causes translational activation. The nature of this activation is not yet understood. From the cases examined here, a picture emerges for RNase III as a versatile element in gene regulation of *E. coli* and its phages.

Acknowledgments. I am very grateful to Dr. D. Court for providing data prior to publication, for his helpful comments and discussions during the preparation of this review, and for his critical reading of the manuscript. I would like to thank Drs. H. Echols, A. Oppenheim, and S. Bear, who consented to the inclusion of their unpublished information in this work. I would like to acknowledge the valuable contributions of my colleague, Dr. Cecilia Montañez, and the skillful typing of the manuscript by Mrs. Patricia Cortés. The author was a John Simon Guggenheim fellow during the development of the research work in his laboratory. These investigations were supported, in part, by a grant from CONACyT (Consejo Nacional de Ciencia y Tecnología), Mexico.

References

Abraham J, Mascarenhas D, Fischer R, Benedik M, Campbell A, Echols H (1980) DNA sequence of regulatory region for integration gene of bacteriophage lambda. Proc Natl Acad Sci USA 77:2477–2481

Abremski K, Gottesman S (1982) Purification of the bacteriophage lambda *xis* gene product required for lambda excisive recombination. J Biol Chem 257:9658–9662

Altuvia S, Locker-Giladi H, Koby S, Ben-Nun O, Oppenheim AB (1987) RNase III stimulates the translation of the *c*III gene of bacteriophage λ. Proc Natl Acad Sci USA 84:6511–6515

Anevski PJ, Lozeron HA (1981) Multiple pathways of RNA processing and decay for the major leftward N-independent RNA transcript of coliphage lambda. Virology 113:39–53

Belasco JG, Beatty JT, Adams CW, von Gabain A, Cohen SN (1985) Differential expression of photosynthesis genes in *R. capsulata* results from segmental differences in stability within the polycistronic *rxc*A transcript. Cell 40:171–181

Belfort M (1980) The *c*II-independent expression of the phage lambda *int* gene in RNase III-defective *E. coli*. Gene 11:149–155

Chung S, Echols H (1977) Positive regulation of integrative recombination by the *c*II and *c*III genes of bacteriophage λ. Virology 79:312–319

Court D, Adhya S, Nash N, Enquist L (1977) The phage λ integration protein (Int) is subject to control by the *c*II and *c*III gene products. In: Bukhari AI, Shapiro JA, Adhya SL (eds) DNA insertion elements, plasmids, and episomes. Cold Spring Harbor Laboratory, New York, pp 389–402

Court D, Huang TF, Oppenheim AB (1983a) Deletion analysis of the retroregulatory site for the λ *int* gene. J Mol Biol 166:233–240

Court D, Schmeissner U, Bear S, Rosenberg M, Oppenheim AB, Montañez C, Guarneros G (1983b) Control of *int* gene expression by RNA processing. In: Hamer D, Rosenberg M (eds) Gene expression. Liss, New York, pp 311–326

Court D, Schmeissner U, Rosenberg M, Oppenheim A, Guarneros G, Montañez C (1983c) Processing of lambda *int* RNA: mechanism for gene control. In: Schlesinger D (ed) Microbiology 1983. American Society for Microbiology, Washington D.C. pp 78–81

Daniels DL, Schroeder JL, Szybalski W, Sanger F, Coulson AR, Hong GF, Hill DF, Petersen GB, Blattner FR (1983) Complete annotated lambda sequence. In: Hendrix RW, Roberts JW, Stahl FW, Weisberg RA (eds) Lambda II. Cold Spring Harbor Laboratory, New York, pp 519–676

Dunn JJ (1976) RNase III cleavage of single stranded RNA. Effect of ionic strength on the fidelity of cleavage. J Biol Chem 251:3807–3814

Dunn JJ, Studier FW (1975) Effect of RNase III cleavage on translation of bacteriophage T7 messenger RNAs. J Mol Biol 99:487–499

Echols H, Guarneros G (1983) Control of integration and excision. In: Hendrix RW, Roberts JW, Stahl FW, Weisberg RA (eds) Lambda II. Cold Spring Harbor Laboratory, New York, pp 75–92

Epp C, Pearson LM, Enquist L (1981) Downstream regulation of *int* gene expression by the *b*2 region in phage lambda. Gene 13:327–337

Fien K, Turck A, Kang I, Kielty S, Wulff DL, McKenney K, Rosenberg M (1984) cII-dependent activation of the *p*RE promoter of coliphage lambda fused to the *Escherichia coli gal*K gene. Gene 32:141–150

Franklin NC (1985) Conservation of genome form but not sequence in the transcription antitermination determinants of bacteriophage λ, φ21 and P22. J Mol Biol 181:75–84

Friedman DI, Gottesman M (1983) Lytic mode of lambda development. In: Hendrix RW, Roberts JW, Stahl FW, Weisberg RA (eds) Lambda II. Cold Spring Harbor Laboratory, New York, pp 21–51

Gottesman M, Oppenheim A, Court D (1982) Retroregulation: control of gene expression from sites distal to the gene. Cell 29:727–728

Guarneros G, Galindo JM (1979) The regulation of integrative recombination by the *b*2 region and the *c*II gene of bacteriophage λ. Virology 95:119–126

Guarneros G, Galindo JM (1984) A post-transcriptional switch for the regulation of bacteriophage lambda *int* gene expression. In: Chopra VL, Joshi BC, Sharma RP, Bansal HC (eds) Genetics: new frontiers. Oxford & IBH Publishing Co, New Delhi, pp 49–58

Guarneros G, Montañez C, Hernandez T, Court D (1982) Post-transcriptional control of bacteriophage λ *int* gene expression from a site distal to the gene. Proc Natl Acad Sci USA 79:238–242

Gupta RS, Kasai T, Schlessinger D (1977) Purification and some novel properties of *Escherichia coli* RNase II. J Biol Chem 252:8945–8949

Hautala JA, Basset CL, Giles NH, Kushner SR (1979) Increased expression of a eukaryotic gene in *Escherichia coli* through stabilization of its messenger RNA. Proc Natl Acad Sci USA 76:5774–5778

Hayashi MN, Hayashi M (1985) Cloned DNA sequences that determine mRNA stability of bacteriophage φX174 in vivo are functional. Nucleic Acids Res 13:5937–5948

Hendrix D (1971) Identification of proteins coded in phage lambda. In: Hershey AD (ed) The bacteriophage lambda. Cold Spring Harbor Laboratory, New York, pp 335–370

Hoess RH, Foeller C, Bidwell K, Landy A (1980) Site-specific recombination functions of bacteriophage lambda: DNA sequence of regulatory regions and overlapping structural genes for Int and Xis. Proc Natl Acad Sci USA 77:2482–2486

Hyman HC, Honigman A (1986) Transcription termination and processing sites in the bacteriophage λ *p*L operon. J Mol Biol 189:131–142

Kasunic DA, Kushner SR (1980) Expression of the *HIS*3 gene of *Saccharomyces cerevisiae* in polynucleotide phosphorylase deficient strains of *Escherichia coli* K12. Gene 12:1–10

Katzir N, Oppenheim A, Belfort M, Oppenheim AB (1976) Activation of the lambda *int* gene by the *c*II and *c*III gene products. Virology 74:324–331

Lehman JF (1974) λ Site-specific recombination: local transcription and an inhibitor specified by the *b*2 region. MGG 130:333–344

Lozeron HA, Dahlberg JE, Szybalski W (1976) Processing of the major leftward mRNA of coliphage lambda. Virology 71:262–277

Lozeron HA, Subbarao MN, Daniels DL, Blattner FR (1983) Transcriptional antitermination and RNase III-mediated processing events of the major RNA transcripts of bacteriophage lambda. Microbiology 1983:74–77

Luk K-C, Dobrzanski P, Szybalski W (1982) Cloning and characterization of the termination site *t*I for the gene *int* transcript in phage lambda. Gene 17:259–262

Mascarenhas D, Kelley R, Campbell A (1981) DNA sequence of the *att* region of coliphage 434. Gene 15:151–156

Mascarenhas D, Trueheart M, Benedik M, Campbell A (1983) Retroregulation: control of integrase expression by the *b*2 region of bacteriophages λ and 434. Virology 124:100–108

McKenney K, Shimatake H, Court D, Schmeissner U, Brady C, Rosenberg M (1981) A system to study promoter and terminator signals recognized by *E. coli* RNA polymerase. In: Chirikjian JC, Papas TS (eds) Gene amplification and analysis, vol II: analysis of nucleic acids by enzymatic methods. Elsevier, New York, pp 383–415

Miller HI, Abraham J, Benedik M, Campbell A, Court D, Echols H, Fischer R, Galindo JM, Guarneros G, Hernandez T, Mascarenhas D, Montañez C, Schindler D, Schmeissner U, Sosa L (1981) Regulation of the integration-excision reaction by bacteriophage lambda. Cold Spring Harbor Symp Quant Biol 45:439–445

Montañez C, Bueno J, Schmeissner U, Court DL, Guarneros G (1986) Mutations of bacteriophage lambda that define independent but overlapping RNA processing and transcription termination sites. J Mol Biol 191:29–37

Mott JE, Galloway JL, Platt T (1985) Maturation of *Escherichia coli* tryptophan operon mRNA: evidence for 3' exonucleolytic processing after rho-dependent termination. EMBO J 4:1887–1891

Nash HA (1975a) Integrative recombination of bacteriophage lambda in vitro. Proc Natl Acad Sci USA 72:1072–1076

Nash HA (1975b) Integrative recombination in bacteriophage lambda: analysis of recombinant DNA. J Mol Biol 91:501–514

Nashimoto H, Uchida H (1985) DNA sequencing of the *Escherichia coli* ribonuclease III gene and its mutations. MGG 201:25–29

Oppenheim AB, Gottesman S, Gottesman M (1982) Regulation of bacteriophage λ *int* gene expression. J Mol Biol 158:327–346

Panayotatos N, Truong K (1985) Cleavage within an RNase III site can control mRNA stability and protein synthesis in vivo. Nucleic Acids Res 13:2227–2240

Plamann MD, Stauffer GV (1985) Characterization of a *cis*-acting regulatory mutation that maps at the distal end of the *Escherichia coli gly*A gene. J Bacteriol 161:650–654

Platt T (1986) Transcription termination and the regulation of gene expression. Ann Rev Biochem 55:339–372

Robertson HD (1982) *Escherichia coli* ribonuclease III cleavage sites. Cell 30:669–672

Robertson HD, Dickson E, Dunn JJ (1977) A nucleotide sequence from ribonuclease III processing site in bacteriophage T7 RNA. Proc Natl Acad Sci USA 74:822–826

Roehrdanz RL, Dove WF (1977) A factor in the *b*2 region affecting site-specific recombinations in lambda. Virology 79:40–49

Rosenberg M, Court D (1979) Regulatory sequences involved in the promotion and termination of RNA transcription. Annu Rev Genet 13:319–353

Rosenberg M, Schmeissner U (1982) Regulation of gene expression by transcription termination and RNA processing. In: Safer B, Grunberg-Manago M (eds) Interaction of transcriptional and translational controls in the regulation of gene expression. Elsevier, New York, pp 1–16

Saito H, Richardson C (1981) Processing of mRNA by ribonuclease III regulates expression of gene 1.2 of bacteriophage T7. Cell 27:533–542

Saito H, Tabor S, Tamanoi F, Richardson CC (1980) Nucleotide sequence of the primary origin of bacteriophage T7 DNA replication: relationship to adjacent genes and regulatory elements. Proc Natl Acad Sci USA 77:3917–3921

Schindler D, Echols H (1981) Retroregulation of the *int* gene of bacteriophage λ. Control of translation completion. Proc Natl Acad Sci USA 78:4475–4479

Schmeissner U, McKenney K, Rosenberg M, Court D (1984a) Removal of terminator structure by RNA processing regulates *int* gene expression. J Biol Mol 176:39–53

Schmeissner U, McKenney K, Rosenberg M, Court D (1984b) Transcription terminator involved in the expression of the *int* gene of phage lambda. Gene 28:343–350

Steitz JA, Bryan RA (1977) Two ribosome binding sites from the gene 0.3 messenger RNA of bacteriophage T7. J Mol Biol 114:527–543

Studier FW, Rosenberg AH, Simon MN, Dunn JJ (1979) Genetic and physical mapping in the early region of bacteriophage T7 DNA. J Mol Biol 135:917–937

Susskind M, Youderian P (1983) Bacteriophage P22 antirepressor and its control. In: Hendrix RW, Roberts JW, Stahl FW, Weisberg RA (eds) Lambda II. Cold Spring Harbor Laboratory, New York, pp 347–364

von Gabain A, Belasco JG, Schottel JL, Chang ACY, Cohen SN (1983) Decay of mRNA in *Escherichia coli*: investigation of the fate of specific segments of transcripts. Proc Natl Acad Sci USA 80:653–657

Weisberg RA, Landy A (1983) Site specific recombination in phage lambda. In: Hendrix RW, Roberts JW, Stahl FW, Weisberg RA (eds) Lambda II. Cold Spring Harbor Laboratory, New York, pp 211–250

Wilder DA, Lozeron HA (1979) Differential modes of processing and decay for the major N-dependent RNA transcripts of coliphage λ. Virology 99:241–256

Wong HC, Chang S (1986) Identification of a positive retroregulator that stabilizes mRNAs in bacteria. Proc Natl Acad Sci USA 83:3233–3237

Wulff DL, Rosenberg M (1983) Establishment of repressor synthesis. In: Hendrix RW, Roberts JW, Stahl FW, Weisberg RA (eds) Lambda II. Cold Spring Harbor Laboratory, New York, pp 53–73

Zaret KS, Sherman F (1984) Mutationally altered 3′ ends of yeast *CYC1* mRNA affect transcript stability and translational efficiency. J Mol Biol 176:107–135

Tail Length Determination in Double-Stranded DNA Bacteriophages

R.W. Hendrix

1 Introduction

The question of how biological structures acquire their specific sizes and shapes is one that can be asked across a wide range, from individual polypeptide chains to redwood trees. Certainly it will turn out that many different mechanisms are involved in the various instances of biological form determination, and as with other large biological questions, it will be necessary to understand many specific cases before it is possible to discern the general themes of how form determination works. This paper discusses one such case, namely length determination in bacteriophage tails. This is a problem for which the main outlines of the solution are now clear, and the more refined details seem experimentally accessible.

2 The Nature of the Problem

The phages in question are the double-stranded (ds) DNA phages with extended tails, for which the coliphages λ and T4 are usually considered the prototypes. These phages have an elaborate structure at the tip of the tail, known as the baseplate in T4 and the basal structure or tail tip in λ, to which is attached the main body of the tail, known as the shaft or tube (Fig. 1). In T4 an additional structure, the sheath, is arranged concentrically around the tube. The tube, for both λ and T4, is a polymer of a single polypeptide subunit (KING 1968; KING and MYKOLAJEWYCZ 1973; BUCHWALD et al. 1970; CASJENS and HENDRIX

Department of Biological Sciences, University of Pittsburgh, Pittsburgh, PA 15260, USA

Current Topics in Microbiology and Immunology, Vol. 136
© Springer-Verlag Berlin·Heidelberg 1988

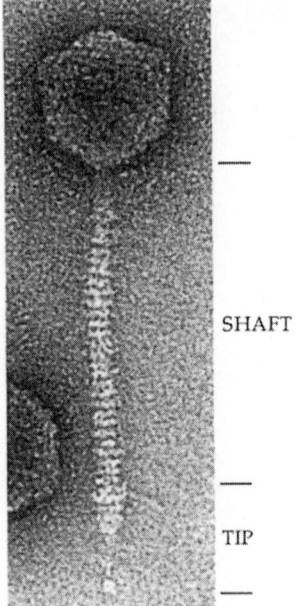

SHAFT

TIP

Fig. 1. Electron micrograph of virions of bacteriophage λ, showing the parts of the tail discussed in the text

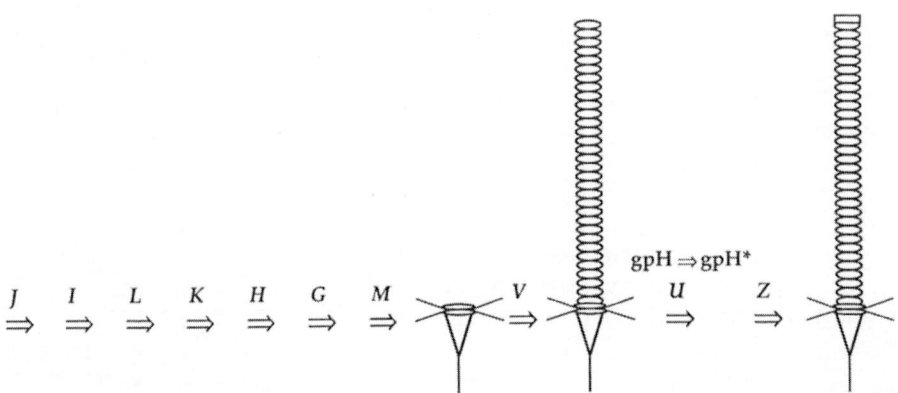

Fig. 2. Assembly pathway for bacteriophage λ tails, adapted from KATSURA and KÜHL (1975b) and supplemented with information from TSUI and HENDRIX (1983). The gene names indicate when those genes have been determined to act in tail assembly. In most cases, "action of a gene" is probably equivalent to addition of the protein encoded by that gene to the nascent tail structure. The relative order of addition of the gpU protein to the tail and cleavage of gpH to gpH* has not been determined

1974), and the problem of understanding how tail length is determined is to understand how the extent of polymerization of this subunit is regulated.

The overall features of the assembly pathways for these phage tails have been worked out, and for the purposes of this discussion the λ and T4 pathways are essentially the same. The pathway for λ tails is shown in Fig. 2. Assembly

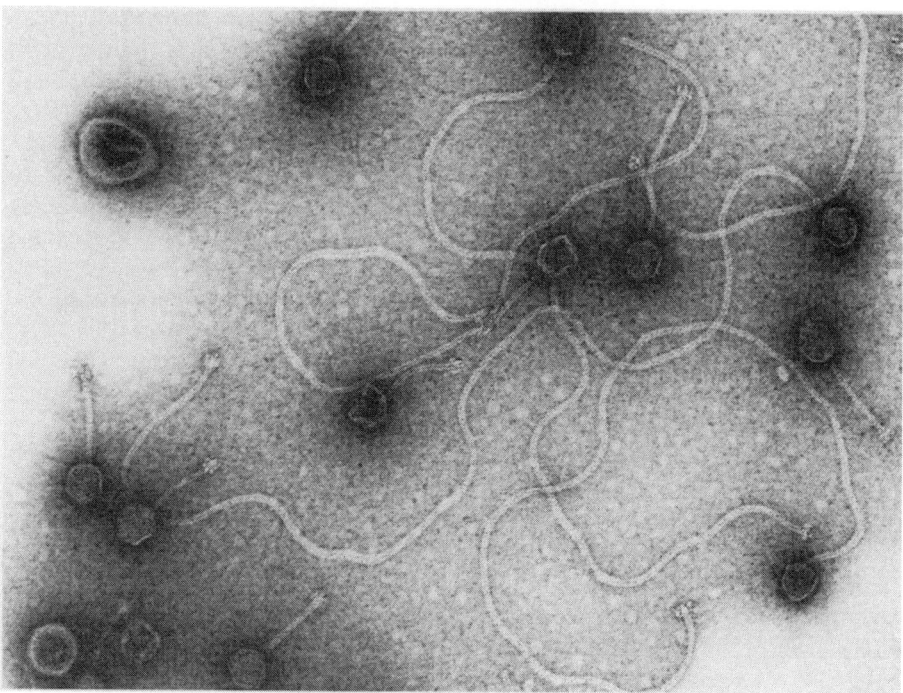

Fig. 3. Virions of bacteriophage MB78 with long tails. A lysate of cells infected by wild-type MB78, which contains mostly phages with normal length tails and about 3% of particles with longer than normal tails, was banded to equilibrium in a CsCl gradient. The material shown here is from the low-density shoulder of the phage band and is therefore highly enriched in particles with exceptionally long tails. A few phages with normal length tails are also present

starts with a series of steps that lead to the formation of the tail tip. The tip then acts as an initiator for gpV, the major shaft subunit, which assembles onto it to form the shaft. Finally, a few additional steps take place which stabilize the structure and render it capable of joining to the head.

The assembly of the shaft is a strikingly precise process. As nearly as can be determined by electron microscopic measurements, the shafts in a population of wild-type phages are virtually all of exactly the same length. In T4, this length is made up of 144 subunits of tube protein gp19, arranged in 24 sixfold symmetric rings (MOODY 1971); in λ, the corresponding numbers for gpV are thought to be 192 subunits in 32 sixfold rings (KATSURA 1983). Rarely, particles with longer than normal tails are found in wild-type populations (an estimated frequency for λ is 10^{-3} to 10^{-4}), and very long tails can be caused to assemble by using certain combinations of tail gene mutations (KATSURA and KÜHL 1975a; LENGYEL et al. 1974; PARKER and EISERLING 1983). One phage, MB78, is known to have the unusually high frequency of 3%–4% long-tailed virions in the wild-type population (JOSI et al. 1982; see Fig. 3). In the cases where long-tailed phages have been tested, their infectivity is diminished from that of normal particles (KATSURA and KÜHL 1975a; HENDRIX, unpublished data).

Three types of models have been considered to explain how shaft length is determined. (For a more detailed discussion of these models than is given here, see Kellenberger 1976.) The first postulates that the subunits of the shaft are themselves intrinsically limited to polymerizing to a particular length (KEL-LENBERGER 1972). WAGENKNECHT and BLOOMFIELD (1975) showed, interestingly, that with reasonable assumptions about the thermodynamic properties of the subunits, such a model could account for the precision in length that is seen. However, the observation that the shaft subunit is capable of polymerizing into a much longer structure than the normal tail, as in the examples cited above, argues against any simple version of this model as an explanation for phage tail length determination.

A second model, the "vernier model," postulates a second structure which co-polymerizes with the tail shaft (ANDERSON and STEPHENS 1964). This second structure is imagined to have a different repeat length than the tail shaft, so that the correct tail length can be signaled when the two structures reach a particular phase relationship. This model seems unlikely because the only known plausible candidate for the second co-polymerizing structure, the sheath of the T4 tail, is clearly not acting in this capacity: the T4 shaft will polymerize efficiently to the correct length in the complete absence of the sheath subunit (KING 1968), and the sheath does *not* have a different repeat period from the shaft (each of the 144 sheath subunits binds identically to one of the 144 shaft subunits) (DEROSIER and KLUG 1968; MOODY 1971). The phage MB78 cited above also provides an argument against the vernier model since the length distribution of the abnormally long tails is smooth (HENDRIX, unpublished data) rather than showing periodic preferred lengths, as might be predicted if a vernier mechanism were operating.

The third model is the "template model". First proposed by KING (1968, 1971) for T4, this model postulates that a protein, acting as a template or tape measure, spans the length of the tail shaft and in some way indicates the point where polymerization is to stop. There is now good reason to believe that the template model is correct, with most of the evidence deriving from experiments with λ and its relatives.

3 The λ Tape Measure Protein

The first strong indication that λ tail length might be determined by a tape measure protein came from experiments by YOUDERIAN and KING (YOUDERIAN 1978; YOUDERIAN and KING, personal communication) comparing the tail genes of λ and $\phi 80$. The tail genes of these two phages are very similar and in some cases are functionally interchangeable. It is therefore possible, by constructing hybrid phages and using other genetic techniques, to produce phages whose tails have been specified by various combinations of λ and $\phi 80$ genes. Since the tail of $\phi 80$ is $\sim 17\%$ longer than λ's tail, their lengths could be measured and correlated with the source of the tail genes. The experiments showed that tail length was determined by a section of DNA containing four tail genes, T, H, M, and L (or their $\phi 80$ equivalents).

The first important conclusion from these results was that the information about how long the tail will be must reside in the proteins that make up the tail tip (a group to which the proteins encoded by those genes belong) and not in the major tail shaft subunit, gpV, or other proteins involved in later assembly steps. Secondly, these results supplied a candidate for a tape measure protein in the product of gene H, gpH. Of the four proteins encoded by the identified genes, only gpH is big enough (90 K) to span the full length of the tail shaft. Gratifyingly, the ratio of molecular weights of λ's gpH and the equivalent $\phi 80$ protein, as measured by SDS gel electrophoresis, is the same as the ratio of tail lengths for the two phages.

These experiments were extended by POPA and HENDRIX (unpublished data), who added three more phages with λ-like morphology but different tail lengths to YOUDERIAN and KING'S comparison of λ and $\phi 80$. These include phage T5 (127% of λ tail length) and two additional lambdoid phages (DHILLON et al. 1980, 1981), HK97 (118%) and HK022 (89%). The virions of all these phages contain proteins in their tails which are presumed to be analogous to λ's gpH because they are present in the same number of copies (~ 6/tail), and like gpH, they undergo a proteolytic cleavage at some time during tail assembly in which they lose ~ 100 amino acids (HENDRIX and CASJENS 1974; YOUDERIAN 1978; ZWEIG and CUMMINGS 1973; POPA and HENDRIX, unpublished data). In agreement with the tape measure hypothesis, the molecular weights of the proteins of all five phages show a roughly linear proportionality to the corresponding tail lengths.

If these putative tape measure proteins measure tail length by comparable mechanisms, then the proportionality constant relating tail length to protein molecular weight should give an indication of how extended the tape measure is when it is measuring. In fact, the constant determined for these phages, 1.5 Å of tail length per amino acid residue of protein, suggests that the tape measure proteins are some approximation to extended α-helices. This supposition is given support by the largely α-helical secondary structure prediction for λ's gpH.

These ideas about tail length determination were put on a more rigorous foundation by KATSURA and HENDRIX (1984), who compared the properties of wild-type λ and two λ mutants that carry in-frame deletions of 219 and 246 bp in gene H. The mutants, which differ from wild type only in the H deletions, make shorter gpH proteins and correspondingly shorter tails. This establishes directly that gpH has a central role in tail length determination. The larger of the two deletions has a larger effect on tail length, a fact that is most easily explained if gpH is indeed directly measuring tail length. Although the effects of these deletions on tail length were small, the measured differences were again consistent with the idea that the tape measure acts in the form of an α-helix. KATSURA and HENDRIX examined the amino-acid sequence of gpH in the central portion of the protein (in which the deletions lie) and found that if the sequence is arranged as an α-helix, hydrophobic residues lie in discernable stripes along the helix. This is characteristic of α-helices that form a coiled

coil structure (CRICK 1953; MCLACHLAN and STEWART 1975) and suggests that the six copies of gpH in the tail may be arranged as a six-stranded coil of α-helices extending the length of the tail tube.

Recently, KATSURA (personal communication) isolated an additional series of 12 in-frame deletions in *H* and 1 internal duplication. These variants produce a wide range of sizes of mutant gpH's – the largest deletion removes 60% of the protein – and a correspondingly wide range of tail sizes. Again, the constant relating protein molecular weight to tail length agrees with the hypothesis that the tape measure is α-helical.

These various lines of experiments all establish that tail length is measured by gpH and strongly imply that it does so by extending along the tail in a largely α-helical form. It remains to be discovered by what mechanism it stops gpV polymerization at the correct position, but a few additional observations may be pertinent. First, it is known at which point in tail assembly gpH adds to the assembling tail tip (KATSURA and KÜHL 1975b; TSUI and HENDRIX 1983), and the change in sedimentation coefficient upon gpH addition is consistent with the hypothesis that the six gpH molecules are in a compact conformation and inconsistent with their being extended. This implies that sometime subsequent to the initial assembly of gpH onto the growing structure, it undergoes a dramatic conformational change, from the compact, initially assembled form into the extended form that we assume it must take to measure tail length. The picture of tail shaft assembly suggested by these observations is that as gpV (shaft) subunits polymerize around the gpH molecules, they force the gpH to unroll into its extended conformation.

Another interesting feature of the biology of gpH which may be important in understanding its role in tail assembly is the proteolytic cleavage mentioned above. GpH is assembled in its uncleaved form (TSUI and HENDRIX 1983), and the cleavage, which removes ~100 amino acids from the C-terminus of the protein (HENDRIX and CASJENS 1974; WALKER et al. 1982), occurs immediately after shaft polymerization is complete (TSUI and HENDRIX 1983). At about the same time, a hexamer of the tail protein gpU adds to the top of the shaft and apparently stabilizes the structure (KATSURA and TSUGITA 1977). One model for the mechanism of length determination that has been discussed proposes that the C-terminal part of gpH is a globular "button" on the end of an α-helical "string" and that the button prevents the shaft from growing any longer once it has polymerized around the string to the length of the string. The cleavage, in this view, would remove the button after it had performed its function, in order to make room for the addition of gpU. Regardless of the correctness of this model, the C-terminal part of gpH also appears to have some role in *promoting* the polymerization of gpV into the shaft: an amber mutation located ~40 codons from the end of gene *H* makes a stable fragment that is longer than the normal cleaved form of gpH. In cells infected with this mutant, gpV polymerization fails to occur to any discernable extent (HENDRIX, unpublished data).

The λ tape measure protein also has another role in addition to its participation in regulating gpV polymerization. Both genetic and biochemical experiments argue that gpH participates in the injection of DNA into the host cell,

and it appears likely that the gpH either enters the cell along with the DNA or aids the DNA in crossing the cell membrane (SCANDELLA and ARBER 1976; ELLIOT and ARBER 1978; ROSSENER and IHLER 1984). The central portion of the polypeptide is important in this process, since all the gene *H* deletion mutants described above are defective in injection to some extent (KATSURA and HENDRIX 1984; KATSURA, personal communication). In contrast, the C-terminal portion of gpH, which appears to be required for gpV polymerization, has no direct role in injection since it is removed by cleavage before injection occurs.

4 Other Phages

Studies on tail length determination with T4, while not providing direct evidence for a tape measure mechanism, have given valuable structural information relating to the putative tape measure. The discussion of the role of gpH in λ tail assembly given above assumed that the gpH occupies the lumen of the tail tube. In T4, DUDA et al. (1985) have shown by direct measurements of mass distributions in the scanning transmission electron microscope that the center of the tube contains material which is removable by guanidine treatment without disrupting the tube and which could plausibly be six, extended, tape measure proteins. The identity of this material is not clear, but gel electrophoresis shows that purified tubes contain baseplate proteins gp48 and gp29 (DUDA et al. 1986). If gp48 is the T4 tape measure protein, it would need to be considerably more extended than the α-helix proposed for λ (DUDA et al. 1986; PRYCIAK et al. 1986). On the other hand, gp29 is about the same size in relation to the T4 tail as is gpH to the λ tail and so might plausibly be a tape measure protein in the λ mold.

Evidence for material in the lumen of the tail tube of another contractile-tailed phage, SP01, has been obtained by conventional electron microscopy (PARKER and EISERLING 1983). Negative stain fails to penetrate the tail tube lumen in normal SP01 virions, but in the rare long tailed variants in the population, stain does penetrate the lumen near the head. Stain penetration stops at the position that would be the top end of a normal tail, suggesting that the lumen below this point is in fact occupied.

5 Prospects

The work described here establishes that the length of phage tails is determined by the length of a tape measure protein. Whether this mechanism will have any generality among other biological structures remains to be seen. However, its greatest value probably lies not in establishing the arrangement for this specific case, but in providing an entrée into the detailed ways in which the proteins involved interact with each other as they accomplish tail assembly and length determination. How does gpH "tell" gpV when to stop polymeriz-

ing? How is gpH's dramatic conformational change accomplished? Answers to these sorts of questions should bring us closer to understanding the "behavioral repertoire" available to proteins as they build a structure. Given the rich variety of genetic, biochemical, and structural techniques that can be applied to bacteriophage systems and with a little help from X-ray crystallography, these phage tails are clearly excellent vehicles for elucidating general principles about what proteins can do and how they do it.

Acknowledgment. Work in this laboratory was supported by Research Grant AI12227 from the National Institutes of Health.

References

Anderson TF, Stephens R (1964) Decomposition of T6 bacteriophage in alkaline solutions. Virology 23:113–116

Buchwald M, Steed-Glaister P, Siminovitch L (1970) The morphogenesis of bacteriophage lambda: I. Purification and characterization of heads and tails. Virology 42:375–389

Casjens S, Hendrix R (1974) Locations and amounts of the major structural proteins in bacteriophage lambda. J Miol Biol 88:535–545

Crick FHC (1953) The packing of α-helices: simple coiled coils. Acta Cryst 6:689–697

DeRosier DJ, Klug A (1968) Reconstruction of three dimensional structures from electron micrographs. Nature 217:130

Dhillon KS, Dhillon TS, Lai ANC, Linn S (1980) Host range, immunity and antigenic properties of lambdoid coliphage HK97. J Gen Virol 50:217–220

Dhillon TS, Dhillon KS, Lai ANC (1981) Genetic recombination between phage HK022, λ, and $\phi 80$. Virology 109:198–200

Duda RL, Wall JS, Hainfeld JF, Sweet RM, Eiserling FA (1985) Mass distribution of a probable tail-length-determining protein in bacteriophage T4. Proc Natl Acad Sci USA 82:5550–5554

Duda RL, Gingery M, Eiserling FA (1986) Potential length determiner and DNA injection protein is extruded from bacteriophage T4 tails in vitro. Virology 151:296–314

Elliot J, Arber W (1978) *E. coli* K-12 *pel⁻* mutants, which block λ DNA injection, coincide with *ptsM*, which determines a component of a sugar transport system. MGG 161:1–8

Hendrix RW, Casjens SR (1974) Protein cleavage in bacteriophage lambda tail assembly. Virology 61:156–159

Joshi A, Siddiqi JZ, Rao GRK, Chakravorty M (1982) MB78, a virulent bacteriophage of *Salmonella typhimurium*. J Virol 41:1038–1043

Katsura I (1983) Tail assembly and injection. In: Hendrix R, Roberts J, Stahl F, Weisberg R (eds) Lambda II. Cold Spring Harbor Laboratory, New York, pp 331–346

Katsura I, Hendrix R (1984) Length determination in bacteriophage lambda tails. Cell 39:691–698

Katsura I, Kühl PW (1975a) Morphogenesis of the tail of bacteriophage λ:II. In vitro formation and properties of phage particles with extra long tails. Virology 63:238–251

Katsura I, Kühl PW (1975b) Morphogenesis of the tail of bacteriophage λ: III. Morphogenetic pathway. J Mol Biol 91:257–273

Katsura I, Tsugitsa A (1977) Purification and characterization of the major protein and the termination protein of the bacteriophage lambda tail. Virology 76:129–145

Kellenberger E (1972) Assembly in biological systems. In: Wolstenholme GEW (ed) Polymerization in biological systems, CIBA Symposium. Elsevier, New York, pp 189–206

Kellenberger E (1976) General discussion. Philos Trans R Soc Lond [Biol] 276:27

King J (1968) Assembly of the tail of bacteriophage T4. J Mol Biol 32:231–262

King J (1971) Bacteriophage T4 tail assembly: four steps in core formation. J Mol Biol 58:693–709

King J, Mykolajewycz N (1973) Bacteriophage T4 tail assembly: proteins of the sheath, core, and baseplate. J Mol Biol 75:339–358

Lengyel JA, Goldstein RN, Marsh M, Calendar R (1974) Structure of the bacteriophage P2 tail. Virology 62:161–174

Mc Lachlan AD, Stewart M (1975) Tropomyocin coiled-coil interactions: evidence for an unstaggered structure. J Mol Biol 98:293–304

Moody MF (1971) Application of optical diffraction to helical structures in the bacteriophage tail. Philos Trans R Soc London [Biol] 261:181

Parker ML, Eiserling FA (1983) Bacteriophage SP01 structure and morphogenesis: I. Tail structure and length determination. J Virol 46:239–249

Pryciak PM, Conway JD, Eiserling FA, Eisenberg D (1986) Cylindrical beta structure: a hypothetical protein structure. In: Protein structure, folding, and design. Liss, New York, pp 243–246

Roessner CA, Ihler G (1984) Protease sensitivity of bacteriophage lambda tail proteins gpJ and pH* in complexes with the lambda receptor. J Bacteriol 157:165–170

Scandella D, Arber W (1976) Phage λ DNA injection into *Escherichia coli pel⁻* mutants is restored by mutations in phage genes V and H. Virology 69:206–215

Tsui L, Hendrix R (1983) Proteolytic processing of phage λ tail protein gpH: timing of the cleavage. Virology 125:257–264

Wagenknecht T, Bloomfield V (1975) Equilibrium mechanisms of length regulation in linear protein aggregates. Biopolymers 14:2297

Walker JE, Auffret AD, Carne A, Gurnett A, Hansch P, Hill D, Saraste M (1982) Solid-phase sequence analysis of polypeptides eluted from polyacrylamide gels. Eur J Biochem 123:253–260

Youderian P (1978) Genetic analysis of the length of the tails of lambdoid bacteriophages. Ph.D. Thesis, Massachusetts Institute of Technology, Cambridge, MA

Zweig M, Cummings D (1973) Cleavage of head and tail proteins during bacteriophage T5 assembly: selective host involvement in the cleavage of a tail protein. J Mol Biol 80:505–518

Single-Stranded DNA Phage Origins

P.D. Baas and H.S. Jansz

1 Introduction

Since the rediscovery of bacteriophage ϕX174 in the 1950s by Robert Sinsheimer (1959), the single-stranded DNA phages have been widely used as model systems in molecular biology. ϕX174 can be considered as their "godfather." The single-

Institute of Molecular Biology and Medical Biotechnology, and Laboratory for Physiological Chemistry, University of Utrecht, Padualaan 8, 3584 CH Utrecht, The Netherlands

Current Topics in Microbiology and Immunology, Vol. 136
© Springer-Verlag Berlin·Heidelberg 1988

stranded DNA bacteriophages are divided into two classes based on the morphology of their representatives, either isometric (icosahedral) or filamentous. The isometric phages follow the conventional infection cycle of adsorption, reproduction, and release of progeny particles after lysis of their host, usually *Escherichia coli* C. The other class, the filamentous phages, do not lyse or kill their host cell. The infected cell continues to grow and to divide, while the progeny virions are formed and extruded through the cell membrane in a continuous fashion. Both classes of single-stranded DNA phages contain a circular genome.

All isometric phages code for the same series of 11 proteins. In spite of this resemblance large variations in the length of the genomes and the nucleotide sequences between various members of this group have been observed. The ϕX genome contains 5386 nucleotides (SANGER et al. 1978), G4 (GODSON et al. 1978) has 5577, and St-1 includes approximately 6000 (GODSON 1978; GRINDLEY and GODSON 1978a, b). The genomes of the filamentous phages code for 10 proteins. The best studied representatives of the filamentous phages (f1, fd, and M13) belong to the Ff group. These phages only infect *E. coli* cells carrying the F^+ transmissible sex factor. In contrast to the large variations observed within the isometric group, the nucleotide sequences of the members of the filamentous Ff group are almost identical. The M13 (VAN WEZENBEEK et al. 1980) and the f1 genomes (BECK and ZINK 1981; HILL and PETERSEN 1982) (6407 nucleotides) are one nucleotide shorter than the fd genome (BECK et al. 1978). M13 and f1 are more closely related to each other than M13 and fd (number of base substitutions 52 and 192, respectively). Most of the base substitutions do not result in amino acid changes in the corresponding 10 proteins of the filamentous phages. Recently the DNA sequence of bacteriophage IKe, a filamentous phage which infects *E. coli* cells harboring plasmids of the N incompatibility group, has been reported (PEETERS et al. 1985). The nucleotide sequence of bacteriophage IKe (6883 nucleotides) differs considerably from those of the Ff group (overall homology 55%).

Studies on the mechanism of DNA replication of the single-stranded DNA phages have revealed three different priming systems for the complementary-strand synthesis. Only host proteins are involved in this process. The viral-strand synthesis, however, is initiated by a phage-specific protein, which cleaves the viral strand at a specific place, the origin of viral-strand synthesis. The purpose of this article is to describe and compare the origins of DNA replication of the various single-stranded DNA phages. Data on the origins of DNA replication of the isometric phages ϕX174, S13, G4, α3, St-1, ϕK, U3 and G14 are available. For the filamentous phages the data are restricted to f1, fd, M13, and IKe. In particular the nucleotide specificity and the interaction of the replication proteins with the origin DNA will be described. Other aspects of the reproduction of these phages are reviewed elsewhere (DENHARDT et al. 1978; KORNBERG 1980, 1982; BAAS 1985; ZINDER and HORIUCHI 1985).

2   The DNA Replication Cycle of Single-Stranded DNA Phages

The replication cycle of single-stranded DNA phages can be divided into three stages. First the viral strand DNA is converted into double-stranded replicative form DNA (RF DNA). In the second stage reproduction of RF DNA takes place. In the final stage the viral strand DNA of the progeny virus is derived from RF DNA by asymmetric synthesis.

2.1   Stage I Replication (SS(c) → RF)

After infection the single-stranded, circular, viral DNA is converted into a covalently closed, double-stranded RF DNA. This conversion is accomplished by host proteins; no phage-encoded proteins are required. The various steps in this process are prepriming, priming, chain elongation by DNA polymerase III holoenzyme, removal of the RNA primers, gap filling by DNA polymerase I, and ring closure by DNA ligase. Finally, superhelical turns are introduced in the parental RF DNA by the action of gyrase (KORNBERG 1980, 1982). In the case of the isometric phages the RNA primer is synthesized on the single-stranded viral DNAs coated with single-stranded, DNA-binding protein of *E. coli* (SSB) by *dna*G protein (primase). Primer formation on bacteriophage ϕX174, S13, G14, and U3 DNA also requires the presence of a pre-primosome complex consisting of one molecule each of protein n′, protein n, *dna*B protein, and probably one or more mulecules of protein i, protein n″, and *dna*C protein. In contrast, priming of G4, St-1, α3, and ϕK DNA does not require preprimosomal formation. The filamentous phages use RNA polymerase instead of *dna*G protein for the synthesis of the RNA primer on their SSB-coated viral DNAs. The three different priming systems are summarized in Fig. 1.

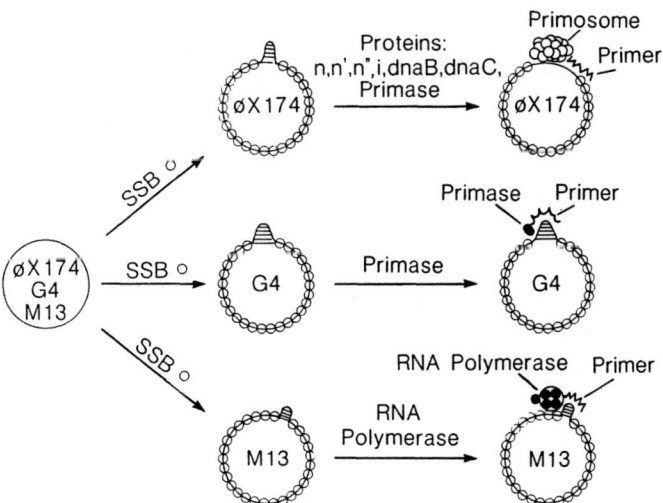

Fig. 1. Scheme for the three different specific priming systems of single-stranded viral DNAs coated with SSB (Adapted from KORNBERG 1980, 1982)

2.2 Stage II Replication (RF → SS(c) → RF)

Replication of the parental RF DNA to yield a pool of progeny RF DNA takes place according to the rolling circle mechanism (GILBERT and DRESSLER 1968). Variances in the method of viral-strand synthesis between isometric and filamentous phages are caused by the different properties of gene A and gene II protein (Table 1). The phage-encoded initiator protein, gene A protein for the isometric and gene II protein for the filamentous phages, cleaves the viral strand of the parental RF DNA at a specific place, the origin of viral-strand synthesis. During cleavage, gene A protein becomes covalently attached to the 5′-phosphate at the cleavage site. The other end of the cleavage site contains a free 3′-hydroxyl group. Together with *rep* protein, gene A protein participates in unwinding of the DNA duplex. DNA polymerase III holoenzyme elongates the viral strand using the 3′-hydroxyl group of the nucleotide at the cleavage site as a primer. During replication, gene A protein still bound to the 5′-end of the tail of the rolling circle travels around the complementary-strand template in advance of replication. In this way a looped rolling circle is formed. At the end of the replication cycle, gene A protein cleaves the regenerated origin, and the 3′- and 5′-ends of the parental viral strand are ligated to form a circle. During this reaction, gene A protein is transferred from the parental to the

Table 1. Comparison of the properties of initiator proteins of isometric and filamentous phages

	Gene A protein	Gene II protein
1. Requirement of supercoiled phage RF DNA for cleavage	yes	yes
2. Secondary structure at cleavage site	low	high
3. Nicking-closing reaction on phage RF DNA	no[a]	yes
4. Cleavage of viral single strands	yes	no
5. Complex formation with phage DNA	attaches covalently to 5′-end of DNA after cleavage	weak
6. Energy for single-stranded DNA circularization from	nicking at start of replication	cleavage after replication
7. Linkage for energy transfer of the enzyme to	5′-end	3′-end
8. Energy conservation during replication cycle	yes	no
9. Mode of action in RF DNA replication	processive	distributive
10. Complementation in vivo	*cis*	*trans*
11. Rolling-circle structures visualized by electron microscopy	looped	extended

[a] Nicking-closing activity of gene A protein has been observed in the presence of Mn^{2+} on ϕX RFI (LANGEVELD et al. 1980) and in the presence of Mg^{2+} on plasmid DNA containing a mutated $\phi X174$ origin (BAAS 1987). The mutated $\phi X174$ origin contains an insertion of seven nucleotides in the spacer region. Adapted from MEYER and GEIDER (1980)

newly synthesized viral strand. This new RFII-gene *A* protein complex can initiate subsequent rounds of DNA replication. The thrown-off circular DNA coated with SSB is converted into a covalently closed, double-stranded RF DNA molecule. The mechanism of complementary-strand synthesis during RF DNA replication mimics stage I replication.

RF DNA replication of the filamentous phages takes place in a similar way, although some differences from the replication cycle of the isometric phages have been observed (Table 1). A major one is the fact that gene II protein is not covalently bound to the 5'-phosphate at the cleavage site but forms a weak complex with the complementary strand opposite the cleavage site. Consequently there are no looped rolling circles, but rolling circles with a loose tail occur as replicative intermediates during the replication cycle of fd RF DNA (MEYER and GEIDER 1982).

2.3 Stage III Replication (RF → SS(c))

During the final stage of the DNA replication cycle, progeny viral strands are synthesized and packaged into phage coats. Viral strand DNA is synthesized in a similar way as during stage II replication. In the case of the isometric phages, complementary-strand synthesis is prevented because the viral strand DNA is replicated directly into a phage prohead. The complementary-strand synthesis of the filamentous phage is repressed by binding of the phage-encoded gene V protein, which replaces SSB on the tail of the rolling circles during stage III replication. Progeny viral strands accumulate late in the infection cycle as DNA-protein complexes. During morphogenesis the gene V protein is replaced by the capsid proteins at the cell membrane.

In conclusion the single-stranded DNA phages utilize during their reproduction two origins of DNA replication, one for the complementary strand and one for the viral strand. The complementary strand origin is recognized by a host protein, and the events leading to the synthesis of viral strand DNA are dictated by the phage-encoded initiator proteins.

3 Localization of the Complementary-Strand Origins

In general, origins of DNA replication have been localized on the bacteriophage genomes by analysis of replicative intermediates isolated from infected cells or by studying *in vitro* replication systems. The development from crude *in vitro* systems to replication systems containing exclusively purified proteins makes it possible to study the initiation reaction of DNA replication in more detail and to localize the exact position of the origin of DNA replication on the genomes. In some cases the first localization of the origin derives from *in vivo* studies. The position of the complementary-strand origins of the filamentous (TABAK et al. 1974) and G4-like phages (BOUCHÉ et al. 1975), however, were first determined in *in vitro* systems. Results obtained from *in vivo* and *in vitro* studies during the last 15 years agree with each other nicely, and to our knowledge and point of view, there are no longer conflicting opinions on the localization of the various origins of DNA replication.

3.1 Bacteriophage $\phi X 174$

The search for the complementary-strand origin of bacteriophage $\phi X 174$ has been a frustrating experience for many investigators, simply because there is no unique starting point of the complementary-strand synthesis. The *in vivo* studies for the determination of the initiation and/or termination sites of the complementary-strand synthesis include both stage I and stage II DNA replication. Bacteriophage $\phi X 174$ DNA replication is restricted to stage I in the presence of 150 µg chloramphenicol/ml, which prevents the synthesis of ϕX gene *A* protein and thereby initiation of stage II replication. Another trick which has been used to study stage I replication is to infect the cells with UV-irradiated phage. The pyrimidine dimers block the forward movement of the replication machinery, and the cells produce only partially duplex circles, e.g., molecules in which the complementary strand has been initiated but not completed. Bacteriophage stage II DNA replication can be studied during infections in the presence of 35 µg chloramphenicol/ml, which prevents synthesis of the $\phi X 174$ proheads needed to initiate stage III replication.

One frequently used technique for the determination of the origin and direction of DNA synthesis was described first by DINTZIS (1961) in his measurement of the direction of synthesis of the hemoglobin protein molecule. DANNA and NATHANS (1972) used this technique for the first time in the field of DNA replication. The rationale behind this technique is as follows. A short pulse label is introduced to an at random-synthesizing population of DNA molecules. If replication starts at a fixed point on the genome and proceeds in an orderly fashion, a gradient of specific radioactivity along the genome should be observed in that part of the molecules that have finished a round of DNA replication during the pulse time. The distribution of the radioactivity along the genome can be studied by restriction enzyme cleavage of such DNA molecules and measurement of the radioactivity of each fragment. The highest and lowest amounts of specific radioactivity then locate the position of the termination and initiation site of DNA synthesis respectively. Data reported by ZUCCARELLI et al. (1976) using this technique during stage I DNA replication do not provide conclusive evidence for a unique initiation site. The highest amount of pulse label was found in gene *A*, suggesting preferential termination of DNA synthesis in this region, but the distribution of the label along the genome did not form a gradient. Similar results were obtained by analysis of pulse-labeled RFI DNA (RF DNA with both strands closed and containing superhelical turns) during stage II DNA replication (BAAS et al. 1978). The complementary strands of these molecules contain the majority of the radioactivity. In another study in which the cells were pulse labeled at 16° C, RFI DNA was isolated containing label in the complementary as well as in the viral strand (GODSON 1974). In that case, the distribution of the label along the $\phi X 174$ genome forms a gradient with the highest amount of radioactivity in gene A. However, the deduced direction of replication from the gradient is that of the viral instead of the complementary strand. Thus, these experiments contribute to the localization of the origin and direction of viral-strand synthesis or overall DNA replication and not to that of the complementary strand.

Another approach that has been used for the determination of the initiation/ termination site of complementary-strand synthesis is the analysis of the discontinuities or gaps in the complementary strand of RF molecules synthesized during stage I and stage II DNA replication. Analysis of partially duplex circles isolated from cells infected with UV-irradiated phage shows that all molecules contain one double-stranded region of variable length and one single-stranded region (BENBOW et al. 1974). Molecules with two single-stranded regions were not detected. The average length of the duplex region is an inverse function of the phage's UV dose. These observations strongly suggest that DNA synthesis is initiated at a single position on each template and that DNA synthesis is stopped when a UV-damaged site is encountered. However, restriction enzyme analysis of these intermediates does not show a unique starting point (HOURCADE and DRESSLER 1978). Also in this study a preferred initiation site near the boundary of the ϕX genes A and H was discovered. Alkaline sucrose gradient centrifugation of RFII DNA (RF DNA with one or more discontinuities in either strand), isolated during stage I replication and containing a nearly full-length complementary strand, showed after linearization with the restriction endonuclease PstI that the discontinuity in the complementary strand did not reside in a unique position on the DNA (BAAS et al. 1978).

Gapped RFII DNA containing a circular viral strand and parts of the complementary strand isolated during stage II DNA replication has been analyzed biochemically (EISENBERG and DENHARDT 1974a, b; EISENBERG et al. 1975) and under the electron microscope (KEEGSTRA et al. 1979). The analysis showed that the gaps were present at many, although not random, locations along the ϕX174 genome. The majority of these molecules contained one gap. The general picture emerging from these *in vivo* studies is that although there exists one initiation site for the complementary-strand synthesis on each template, the position of this initiation site differs from molecule to molecule. This conclusion was confirmed by *in vitro* studies. These studies also showed that the discontinuity in the nearly full-length complementary strand is not located at a unique position on the ϕX174 genome (TABAK et al. 1974). Investigation of the primers made on the ϕX174 genome in the absence of DNA synthesis showed that several RNA primers (5–8) can be made on each circle (MCMACKEN et al. 1977). The size of the RNA primers varies from 10–60 nucleotides. The complexity of the fingerprint pattern of the oligoribonucleotides obtained after T1 RNase digestion of the primers indicates that the primers were synthesized at multiple sites on ϕX174 DNA. This was confirmed by a hybridization study which showed that the primers hybridized with each restriction fragment of ϕX174 RF DNA obtained by digestion with the restriction endonucleases HaeIII, HpaII, AluI, and HhaI (ARAI et al. 1981). Similar results with regard to the location of the primers on the ϕX174 genome were obtained when the primers were synthesized in the presence of DNA polymerase III holoenzyme

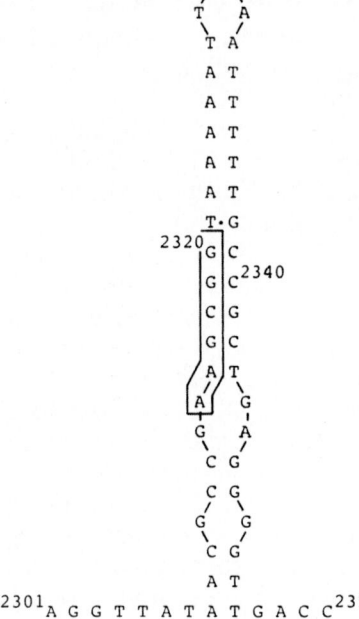

Fig. 2. Possible secondary structure of DNA sequences at the recognition site of protein n′ (Y factor) of bacteriophage ϕX174. *Numbers* indicate nucleotide positions on the ϕX174 map. The consensus sequence A A G C G G found at different n′ recognition sites is *boxed*

(OGAWA et al. 1983). However, in this case the length of the primers is significantly shorter, 4–7 nucleotides, than in the absence of DNA polymerase III holoenzyme. These data suggest, that the primosome functions as a "mobile promoter" which can start RNA synthesis at many different places on the genome. In the presence of DNA polymerase III holoenzyme, the first synthesized primer is rapidly elongated to a full-length, linear, complementary strand. Although the *in vitro* studies also showed an initiation of the complementary-strand synthesis at multiple sites on the ϕX174 genome, the study of the mechanism of primosome formation on SSB-coated ϕX174 DNA revealed that the primosome is assembled at a specific site. Purification of the proteins involved in primosome formation have led to the discovery that protein n′ (protein Y in the nomenclature of WICKNER and HURWITZ 1975) possesses ϕX174 DNA-specific ATPase or dATPase activity and can bind to SSB-coated ϕX174 DNA. Digestion of ϕX174 DNA with restriction enzymes followed by exonuclease VII treatment shows that the recognition sequence of protein n′ (e.g., ATPase activity) lies within a 54-nucleotide fragment, nucleotides 2301-2354 of the ϕX174 DNA sequence (SHLOMAI and KORNBERG 1980a) (Fig. 2). This sequence is located in the intergenic region between the ϕX174 genes *F* and *G*. This fragment contains a 44-nucleotide sequence with a potential hairpin structure, which remains uncoated in the presence of SSB. Primosome-specific DNA replication and (d)ATPase activity only occur when this particular sequence is present on the template (ARAI and KORNBERG 1981). Primosome formation is thought to start with the recognition by protein n' of the hairpin on SSB-coated ϕX174 DNA. After it is complete, the primosome moves processively along the viral circle in the 5' to 3' orientation of the viral strand, a direction opposite to primer synthesis and DNA chain elongation. The observation that protein n' is able to displace SSB bound to the template DNA (SHLOMAI and KORNBERG 1980b) and the results from a comparative study on the effects of ATP, ATP analogs, and dATP on the formation of primers by the general priming system (consisting of *dna*B and primase) and the primosome system lead to a model in which ATP and/or dATP hydrolysis by protein n' provides the energy for primosome movement, while *dna*B is responsible for engineering a DNA region in which primase can function (ARAI et al. 1981).

3.2 G4-like Phages

In vitro studies show that the conversion of the viral strand of bacteriophage G4 into RFII DNA is accomplished by SSB, DNA polymerase III holoenzyme, and *dna*G protein (ZECHEL et al. 1975). Digestion of G4 RFII DNA by *Eco*RI, followed by sucrose gradient centrifugation, shows that the initiation site of the complementary-strand synthesis is located close to the unique *Eco*RI cleavage site in G4 RF DNA (BOUCHÉ et al. 1975). A primer RNA with a length of 26–28 nucleotides was isolated upon incubation of SSB-coated G4 DNA in the presence of the four ribonucleoside triphosphates with primase. RNA sequence analysis of this primer reveals a unique sequence (BOUCHÉ et al. 1978). This sequence is found in the intergenic region between genes *F* and *G* (FIDDES et al. 1978; SIMS and DRESSLER 1978). Thus, although the mechanisms for primer formation on ϕX174 and G4 DNA are completely different, the site responsible for the initiation of complementary-strand synthesis is located at the same position on the chromosome.

In vivo studies have pinpointed the origin of complementary-strand synthesis of G4 in the same region of the genome. These studies include restriction enzyme analysis of the partially duplex DNA molecules isolated from cells infected with UV-irradiated G4 phage (HOURCADE and DRESSLER 1978) and determination of the location of the discontinuity in a nearly full-length complementary strand of RFII DNA isolated during stage II replication. The site is determined through the unique *Eco*RI and *Pst*I restriction sites and by limited nick translation of the complementary strand using DNA polymerase I and α-^{32}P-labeled deoxyribonucleotides followed by restriction enzyme analysis (MARTIN and GODSON 1977).

The isometric bacteriophages St-1, ϕK, and α_3 utilize the same proteins as G4 for the synthesis of their complementary strands. The initiation sites of this synthesis are determined by RNA sequencing of the primer RNAs made *in vitro* and by DNA sequencing of the corresponding region of the genomes (SIMS et al. 1979; BENZ et al. 1980a). Surprisingly, they are located in the intergenic region between genes *G* and *H*. Although some differences have been found, the nucleotide sequences of the origins are well conserved (Fig. 3).

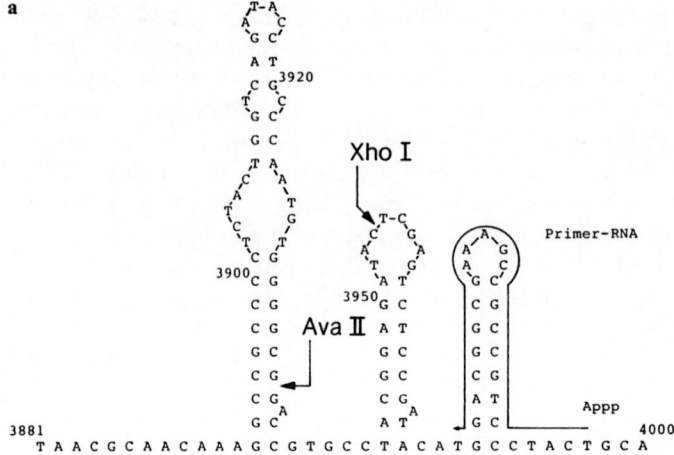

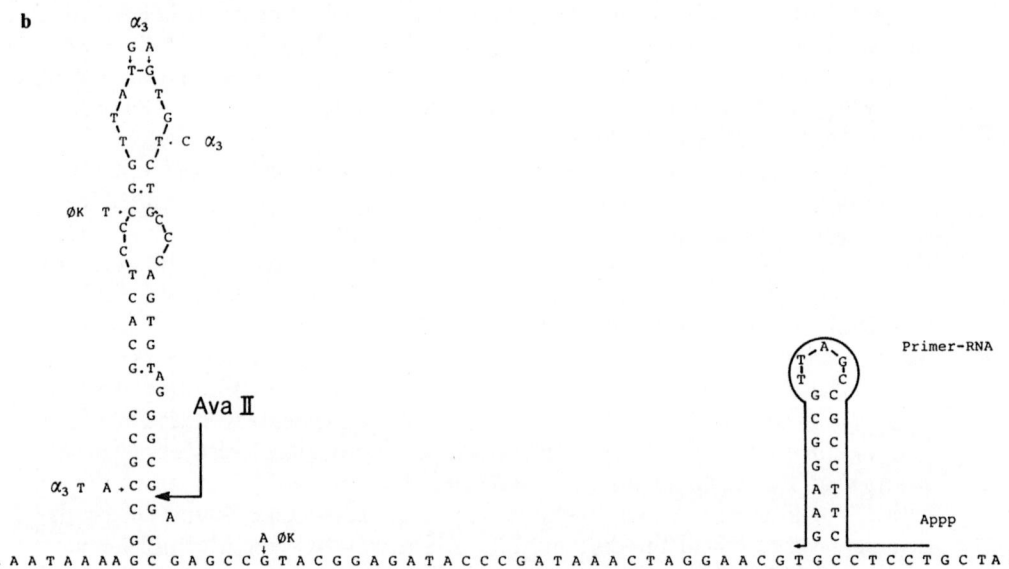

Fig. 3a, b. Possible secondary structure of DNA sequences at the complementary-strand origin of bacteriophage G4 (**a**) and bacteriophages St-1, ϕK, and α_3 (**b**). **a** *Numbers* indicate nucleotide positions on the G4 map. **b** Nucleotide differences for bacteriophages ϕK and α_3 from bacteriophage St-1 are indicated. The position of the primer RNAs and the restriction sites used for mutational studies are indicated (see text)

3.3 Filamentous Phages

The complementary strand of the filamentous phages is initiated by an RNA primer synthesized by RNA polymerase. The discovery of this process both *in vivo* (BRUTLAG et al. 1971) and *in vitro* (WESTERGAARD et al. 1972; WICKNER et al. 1972) is the first example of the involvement of RNA synthesis in the

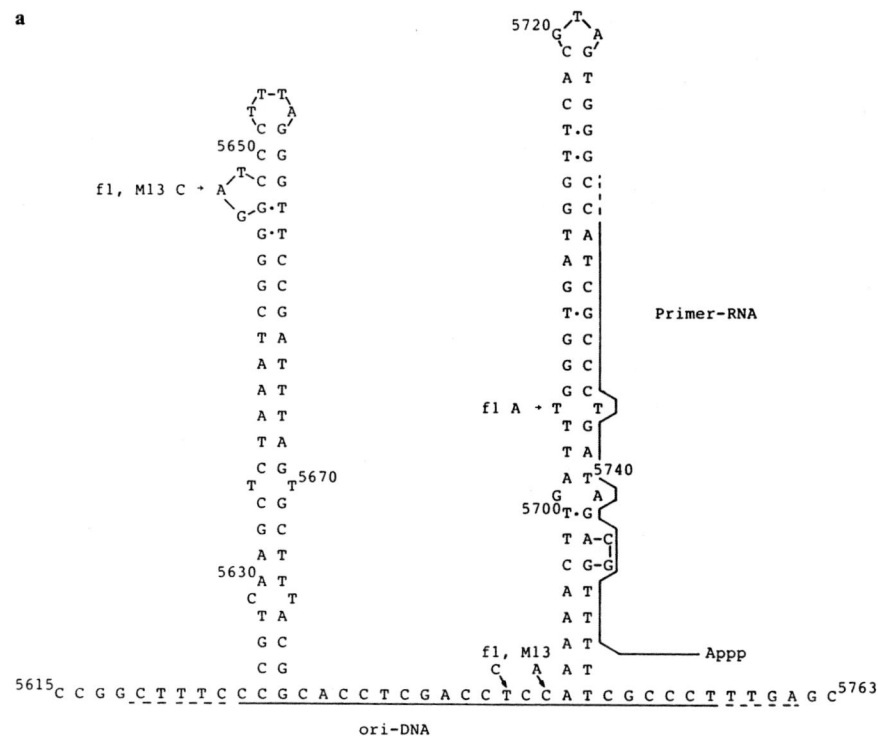

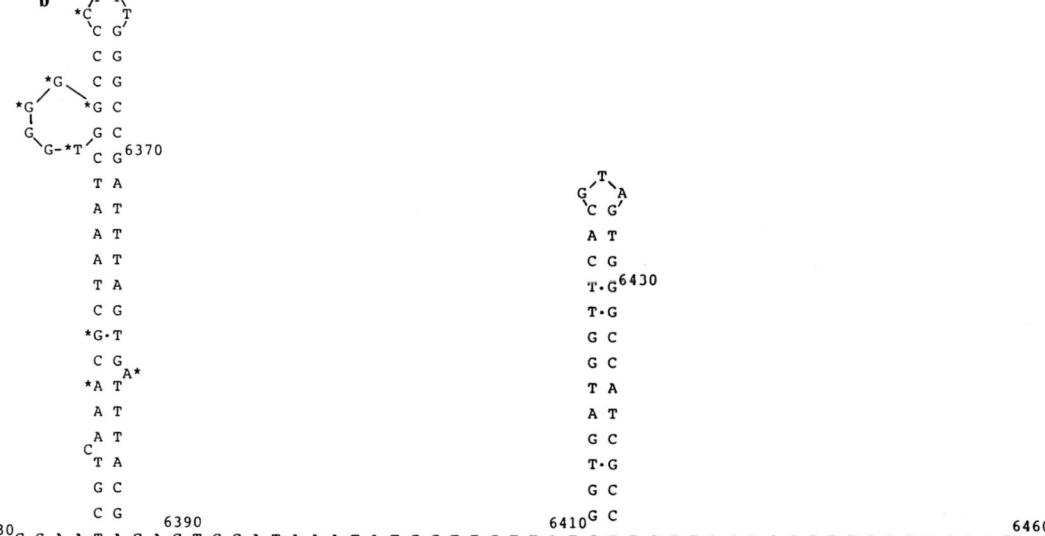

Fig. 4a, b. Possible secondary structure of DNA sequences at the complementary-strand origin of the filamentous phages fd (**a**) and IKe (**b**). *Numbers* indicate nucleotide positions. The locations of the RNA-polymerase-protected fragment (ori-DNA) and the primer RNA of bacteriophage fd are given. Included in **a** are the nucleotide differences between fd, M13, and f1. In the IKe DNA sequence nucleotide differences from the Ff group only in the hairpin are indicated with an *asterisk*

initiation of DNA synthesis. Initiation by RNA polymerase takes place at a unique position on the SSB-coated template. TABAK et al. (1974) show that the gap in the complementary strand of RFII DNA is located close to the unique *Hin*dII restriction site and within the *Hpa*II H restriction fragment. Later, when more data concerning the physical and genetic maps of the filamentous phages became available, the origin of complementary-strand synthesis could be located in the intergenic region between genes II and IV. The exact beginning of complementary-strand synthesis has been determined by RNA sequence analysis of the approximately 30-nucleotide-long RNA primer synthesized by RNA polymerase on SSB-coated fd DNA (GEIDER et al. 1978) (Fig. 4).

Measurement of the distribution of the complementary-strand label in RFI DNA, pulse labeled during stage II replication, shows that *in vivo* complementary strand DNA is initiated at the same site as deduced from the *in vitro* studies (HORIUCHI and ZINDER 1976; HORIUCHI et al. 1979).

4   Nucleotide Sequence Specificity of Complementary-Strand Origins and Interaction with Initiator Proteins

4.1   The Primosome Assembly Sites

The primosome consists solely of host proteins, some of which are clearly required for *E. coli* DNA replication. Therefore, the primosome is the accepted model of priming host lagging strand synthesis. Several investigators have detected in the *E. coli* chromosome (RAY et al. 1981; VAN DER ENDE et al. 1983) or in *E. coli* plasmids, such as pBR 322, Col E1, clo DF13, and mini F, DNA sequences that can function as primosome assembly sites (pas) (ZIPURSKY and MARIANS 1980, 1981; NOMURA et al. 1982a, b; VAN DER ENDE et al. 1983; IMBER et al. 1983). This function was demonstrated either by protein n' ATPase activity or by the ability of these sequences, when cloned into a filamentous phage vector, to confer upon that phage DNA the ability to be converted to RF DNA in a rifampicin-resistant, *dna*B-, *dna*C- and *dna*G-dependent manner. The capability of these sequences, when packaged as part of a single-stranded plasmid DNA molecule in a ϕX174 coat, to enhance the transduction frequency upon infection of ϕX174-sensitive cells was also used to detect n' protein recognition sites. Comparison of the DNA sequences that are recognized by n' protein showed stem-and-loop structures, in which the length of the loop as well as the length of the stem is variable. The total length of the stem-and-loop structures consists of 50 to approximately 70 nucleotides. The nucleotide sequences of the various n' protein recognition sites differ considerably. The n' protein recognition sites on pBR 322 DNA are G-C rich, whereas the n' recognition site on ϕX174 DNA is A-T rich. However, the stem of the different recognition sites has a stretch of six nucleotides in common, 5'-AAGCGG-3', which may be important in the recognition of the DNA by n' protein (MARIANS et al. 1982; IMBER et al. 1983; VAN DER ENDE et al. 1983).

GREENBAUM and MARIANS (1984) have investigated the binding of n' protein to the n' recognition site of ϕX174 (pas-x), as well as to those of pBR322 (pas-BH and pas-BL), using DNase footprinting and methylation enhancement/protection as probes. These studies show that n' protein binds to the entire length of the effector site. No experiments have been reported that determine nucleotide specificity and the minimum size of pas-x. The 54-nucleotide fragment obtained after exonuclease VII digestion of the *Hae*III restriction fragment Z1 possesses approximately 50% of the ATPase activity compared to the whole ϕX174 genome or the Z1 restriction fragment. This lower activity and the fact that three nucleotide residues, which are located in a second small hairpin at the 3' side of pas-x, are protected by n' protein from methylation suggests that for full activity of pas-x also, as in the case of pas-BH and pas-BL, approximately 70 nucleotides are required. Extensive mutational analysis of pas-BH and pas-BL has been carried out in the laboratory of MARIANS (SOELLER and MARIANS 1982; ABARZUA et al.; SOELLER et al. 1983, 1984; ABARZUA and MARIANS 1984). The conclusions of that study are probably also applicable for the n' protein recognition site of ϕX174. Four classes of mutants were obtained, which were characterized for their n' protein ATPase activity as well as their template activity for DNA synthesis using purified proteins. Class 1 mutants are silent mutants which have no effect in any reaction. Class 2 mutants require a higher Mg^{2+} concentration than the wild-type DNA in the ATPase assay. Class 3 mutants are inactive in the ATPase assay and have severely reduced levels of replication template activity (six to seven fold less than the wild-type DNA). Class 4 mutants behave in a manner similar to class 2 mutants for ATPase activity but have a replication template activity intermediate between that of class 2 and class 3 mutant DNAs. The effect of various ligands on the ATPase activity of class 2 mutants and the effect of multiple base substitutions show that for proper functioning primarily a compact, base-paired, tertiary structure of the n' protein recognition site is important. Single-base alterations in the essential sites are often not enough to inactivate completely pas-BL *in vivo* or *in vitro*, but two or more alterations are sufficient for its inactivation. DNA sequence analysis of the revertants and active second-site mutations also strongly suggests that a double-helical, base-paired, DNA stem plays an important role in the recognition of n' protein. All these mutations enhance the stability of the stem structures of pas sites.

4.2 The Primase Recognition Site (G 4-like Phages)

The precise position of the origin of the complementary-strand synthesis of the G4-like phages was determined by sequencing the primer RNA made by primase on the SSB-coated phage DNAs (BOUCHÉ et al. 1978; SIMS et al. 1979; BENZ et al. 1980a). A stretch of approximately 120 nucleotides seems to be required for recognition by one (SIMS and BENZ 1980) or two (STAYTON and KORNBERG 1983) molecules of *dna*G protein. Within this DNA segment two (in St-1, ϕK, and α3) or three (in G4) regions have the potential to form hairpin structures. One of the hairpin structures is the template for the primer

RNA (Fig. 3). Footprinting experiments using various nucleases show that three, well-separated groups of nucleotides within the origin region are protected in the presence of SSB from nuclease digestion by *dna*G protein (BENZ et al. 1980b; SIMS and BENZ 1980). The protected areas are located in the stem and at the base of the primer hairpin, and in the regions flanking the large downstream hairpin. After cleavage of the origin fragment (the seventh largest ϕK *Hpa*II restriction fragment) with the restriction endonuclease *Ava*II, which cleaves at the base of the downstream hairpin, *dna*G protein does not bind to the primer hairpin. This indicates that the downstream hairpin is required for binding of *dna*G protein. This conclusion is confirmed by cloning the separate hairpins into a filamentous phage vector. Using the *Xho*I restriction site present in the loop of the middle hairpin of G4, SAKAI and GODSON (1985) constructed filamentous phages containing either the primer or the downstream hairpin. Both phages were not able to convert their single-stranded phage DNA into RF DNA in the presence of rifampicin, indicating that the primase-dependent origin was not functional. The precise left and right boundary of the G4 complementary-strand origin has not been determined. Insertion mutagenesis using the *Xho*I restriction site showed that several hundreds of nucleotides can be inserted between the primer and the downstream hairpin without loss of origin function (SAKAI and GODSON 1985). Probably because of the middle hairpin, the spatial distance between the primer and the downstream hairpin is not changed so much that complex formation with one or two primase molecules is prohibited. Insertion of nucleotides at the *Ava*II restriction site, however, impairs the origin function to a greater or lesser extent, indicating the importance of nucleotides at the base of the downstream hairpin and/or the distance between the two hairpins (SAKAI et al. 1985, 1987; LAMBERT et al. 1986).

Single base substitutions in the primer hairpin at positions 3976 (G → C) and 3988 (C → G) of the G_4 DNA sequence cause a temperature dependent reduction in origin function *in vivo*. The double mutant in which the hairpin structure is restored, shows a slight temperature independent reduction in origin function (LAMBERT et al. 1987). These experiments indicate, as in the case of the n' protein recognition site, that a secondary DNA structure as well as a specific DNA sequence are required for proper interaction of primase with the G_4 complementary strand origin.

4.3 The Filamentous Phage Complementary-Strand Origin

RNA polymerase protects in the presence of SSB a region of approximately 125 nucleotides of the viral strand of the filamentous phages from nuclease digestion (SCHALLER et al. 1976). The protected region has been called ori-DNA. The nucleotide sequence of this ori-DNA can be folded by base pairing of self-complementary stretches of bases into two, large, hairpin structures (GRAY et al. 1978). One of these hairpins is used as a template for the synthesis of a primer RNA of approximately 30 nucleotides (Fig. 4). As in the case of the G4-like phages, the exact position of the complementary-strand origin was determined by sequence analysis of the primer RNA synthesized by RNA polymerase on SSB-coated fd RNA (GEIDER et al. 1978). The minimum DNA sequence required for RNA polymerase-dependent initiation of the filamentous complementary-strand synthesis has not been determined. Also, no site-directed mutagenesis of the origin region to establish essential nucleotides has been reported. Comparison of the nucleotide sequences of the nonhomologous filamentous phages Ff and IKe shows strong conservation of the nucleotide sequence in the two hairpin structures (Fig. 4). KIM et al. (1981) have introduced deletions into the complementary-strand origin by exploiting the single *Asu*I restriction site located close to the RNA-DNA junction of the M 13 complementary strand. Mutants lacking the primer or both hairpin structures form faint plaques, give reduced phage yields, and show a lag in phage production. Apparently, the RNA polymerase-protected hairpins and the RNA primer coding sequence are important, but not essential, for replication. SOELLER et al. (1983) have demonstrated that these deletion mutants replicate *in vitro* by a ϕX174 primosome-dependent pathway. Because of their reduced growth and plaque morphology, these deletion mutants have been used for the selection of DNA sequences capable of promoting DNA chain initiation and for mutational studies of complementary-strand origins.

5 Localization of the Viral-Strand Origins

5.1 Bacteriophage ϕX174

The first paper about the localization of the origin of ϕX174 DNA replication described genetic experiments in which the progeny phage from spheroplasts infected with heteroduplex ϕX174 RF DNA was analyzed (BAAS and JANSZ 1972). A fraction of the spheroplasts produced a mixed burst of both genotypes. A gradient in the percentage of mixed bursts was observed depending on the position of the heteroduplex marker on the genetic map. These results could be explained by assuming a competition between the heteroduplex DNA repair system and DNA replication. In order to produce a mixed burst, the heteroduplex must avoid the repair system of the host by replication. Therefore, a heteroduplex marker located just behind the origin of replication has a better chance of escaping repair than a heteroduplex located near the terminus of DNA replication. The gradient in the percentage of mixed bursts suggested that the origin of DNA replication is located in the N-terminal part of gene *A* and that DNA replication proceeds unidirectionally in the direction of the genes *A* through *H*.

Further *in vivo* studies concerning the initiation site of viral-strand synthesis have concentrated on the position of the discontinuity in the viral strand of RF molecules, the distribution of the radioactivity along the viral strand in pulse-labeled RF DNA, and the analysis of rolling circle intermediates under the electron microscope. These various analyses include both stage II and stage III replication.

EISENBERG et al. (1975) have shown that RFII DNA isolated during stage II replication contains a specific gap in the viral strand of the *Hin*dII restriction fragment R 3. Pyrimidine tract analysis of the filled-in gaps shows a fourfold enrichment for the unique C_6T tract as compared with the whole viral strand. The C_6T tract is located in the *Hae*III restriction fragment Z6B, 13 nucleotides upstream from the position of the gene *A* protein cleavage site (LANGEVELD et al. 1978). Similar experiments performed by JOHNSON and SINSHEIMER (1974) using RFII DNA isolated during stage III replication show a specific discontinuity of the viral strand in the *Hind*II restriction fragment R 3.

   ϕX174 DNA replication has been studied extensively in *rep*⁻ cells. In these cells ϕX174 RF DNA is formed, but further DNA replication is blocked because of lack of the *rep* gene product. FRANCKE and RAY (1971, 1972) show that after transcription and translation the viral strand of the parental RF DNA is specifically nicked if a functional gene *A* protein is present. Therefore, this system seems extremely useful for studying the location of the ϕX174 gene *A* protein cleavage site *in vivo*. However, further analysis shows that this discontinuity does not reside in a specific location on the ϕX174 chromosome. BAAS et al. (1976) and BOWMAN and RAY (1975) provide evidence that, after the introduction of the nick, extensive degradation of the parental viral strand takes place. This degradation is presumably caused by a nick translation process, starting at the nick introduced by the gene *A* protein. Using this nick translation process, the original position of the ϕX174 gene *A* protein cleavage site could be determined in the *Hae*III restriction fragment Z6B at approximately 100 nucleotides from the Z2/Z6B junction (BAAS et al. 1976) which corresponds closely to the position of the gene *A* protein cleavage site as determined later by *in vitro* experiments (LANGEVELD et al. 1978). Determination of the distribution of the radioactivity along the viral strand in RFI and RFII molecules isolated after pulse-labeling of ϕX174-infected cells during stage II and/or stage III replication confirms the position of the origin and direction of viral-strand replication (GODSON 1974; BAAS et al. 1978).

   KOTHS and DRESSLER (1978, 1980) and KEEGSTRA et al. (1979) have analyzed ϕX174 replicative intermediates under the electron microscope. The position of the ϕX174 gene *A* protein cleavage site was determined in these studies by analysis of the rolling circle intermediates after cleavage with the restriction enzyme *Pst*I, which linearizes ϕX174 RF DNA, or after annealing a specific restriction fragment to the tail of the rolling circle. The results indicated that the initiation site of viral-strand synthesis is located in the *Hae*III restriction fragment Z6B.

   As in the case of the complementary-strand origins, the exact localization of the viral-strand origin becomes possible through *in vitro* studies using the purified initiator, gene *A* protein. After incubation of superhelical ϕX174 RFI DNA with gene *A* protein, the gene *A* protein cleavage site is determined following digestion of the nicked DNA with restriction endonucleases and analysis of the reaction products by gel electrophoresis, electron microscopy, and DNA sequencing. After cleavage of the DNA the gene *A* protein is covalently attached to the DNA at the 5′-end of the cleavage site. This covalent linkage prevents the DNA restriction fragment containing the gene *A* protein (the *Hind*II R3 fragment) from entering the gel (IKEDA et al. 1976). Under the electron microscope the gene *A* protein is visualized as a knob on *Pst*I-linearized ϕX174

DNA at a distance of 20% to the nearer end of the molecule (EISENBERG et al. 1977). LANGEVELD et al. (1978) determined the nucleotide sequence at the 3'-end of the cleavage site. They isolated the two parts of the cleaved viral strand of the *Hae*III restriction fragment Z6B by electrophoresis under denaturing conditions. Labeling of the 3'-OH group at the cleavage site by terminal transferase followed by nucleotide sequence analysis shows a unique sequence 5'-TGCTCCCCCAACTTG-OH-3', corresponding to nucleotides 4291–4305 of the ϕX174 DNA sequence (SANGER et al. 1978).

5.2 Bacteriophage G4

In principle the same techniques which are used for the determination of the viral-strand origin of ϕX174 have also been applied to G4. RAY and DUEBER (1975) have determined the discontinuity in the viral strand of G4 RFII DNA isolated from rep^--infected cells and of G4 RFII DNA isolated during stage III replication. Similar results are obtained in both experiments. The discontinuity in the viral strand is located opposite the unique *Eco*RI restriction site in the *Hae*III Z2A restriction fragment. This is demonstrated by alkaline sucrose gradient centrifugation of G4 RFII DNA after digestion with the restriction endonuclease *Eco*RI. Apparently the nick translation process in the rep^- cells does not proceed very far in this particular experiment. Restriction enzyme analysis after limited enzymatic repair by DNA polymerase I of RFII molecules isolated during single-stranded DNA synthesis with α-^{32}P-labeled deoxyribonucleoside triphosphates also localizes the discontinuity in the viral strand to the *Hae*III restriction fragment Z2A. Similar results are obtained by MARTIN and GODSON (1977). Analysis of the rolling circle intermediates under the electron microscope confirms the position of the viral-strand origin (GODSON 1977).

The exact position of the viral strand origin was determined after cleavage *in vitro* of superhelical G4 RFI DNA with gene *A* protein. ϕX174 gene *A* protein (VAN MANSFELD et al. 1979) as well as G4 gene *A* protein (WEISBEEK et al. 1981) cleaves the viral strand of G4 RFI DNA at the same site between nucleotides (506G) and (507A) of the G4 DNA sequence (GODSON et al. 1978). This position is established by DNA sequence analysis of the cleaved viral strand of the *Taq*I 10 restriction fragment.

5.3 Filamentous Phages

The origin of viral-strand replication of bacteriophage f1 has been determined during stage II and stage III DNA replication by HORIUCHI and ZINDER (1976) by measuring the distribution of radioactivity along RFI DNA after a short pulse of [^{3}H]thymidine. A gradient in the radioactivity is observed, indicating that viral strand synthesis starts in the *Hae*III G restriction fragment, which is located in the intergenic region between genes II and IV on the genetic map. Similar results are obtained by SUGGS and RAY (1977) with pulse-labeled RFI DNA isolated during stage III replication from M13-infected cells. In addition,

SUGGS and RAY (1977) also localize the discontinuity in the viral strand of late M13 RFII DNA at approximately 10% from the unique *Hin*dII cleavage site by alkaline sucrose gradient centrifugation. The position of the gap in late RFII DNA is confirmed by limited repair synthesis by *E. coli* DNA polymerase I using α-^{32}P-labeled deoxyribonucleoside triphosphates followed by restriction enzyme analysis. Partial denaturation mapping of rolling circle intermediates isolated late in M13 infection under the electron microscope and analysis of those replicative intermediates after cleavage with the restriction endonuclease *Hin*dII also place the origin of viral-strand synthesis in the intergenic region between genes II and IV (ALLISON et al. 1977).

The cleavage site of gene II protein and therefore the exact position of the origin of viral-strand synthesis has been determined by MEYER et al. (1979). In a similar way as described above for the ϕX174 gene *A* protein cleavage site, the two parts of the cleaved viral strand of the *Hae*III restriction fragment G are isolated and labeled with terminal transferase, followed by DNA sequence analysis. The sequence analysis shows that gene II protein cleaves the phosphodiester bond between nucleotides 5781 (T) and 5782 (A) of the fd DNA sequence. NOMURA and RAY (1980) determine in a similar way the cleavage site of gene II protein in RFII DNA isolated late in infection from an *E. coli* strain defective in the 5' to 3' exonuclease associated with DNA polymerase I.

6 Nucleotide Sequence Specificity of Viral-Strand Origins and Interaction with Initiator Proteins

6.1 The Isometric Phage Viral-Strand Origin

Comparative studies on the origins of viral-strand synthesis have revealed that a stretch of 30 nucleotides located around the gene *A* protein cleavage site is highly conserved in bacteriophage ϕX174, S13, G4, G14, and U3 DNA (SANGER et al. 1978; LAU and SPENCER 1985; GODSON et al. 1978; HEIDEKAMP et al. 1982). In the isometric phages α3 and St-1 only two nucleotide changes within this region have been detected (HEIDEKAMP et al. 1980, 1982); in the other phages this region remains unchanged. Outside the origin region many nucleotide differences exist between the various members of the isometric phage group. Several of these phage RF DNAs have been used as a template in the ϕX174 RF→SS(c) *in vitro* DNA replication system (DUGUET et al. 1979; KORNBERG 1980, 1982), and from them (except S13) the ϕX174 gene *A* protein cleavage site has been determined by DNA sequence analysis after isolation of the cleaved viral strand (LANGEVELD et al. 1978; VAN MANSFELD et al. 1979; HEIDEKAMP et al. 1980, 1982). In all cases the ϕX174 gene *A* protein cleaves the phosphodiester bond between the G and A residues in positions 7 and 8 of the origin region (Fig. 5).

The conservation of the 30-nucleotide origin region during evolution of the isometric phages strongly suggests that for the reproduction of these phages, a region of approximately 30 nucleotides is important. The fact that within

Fig. 5. The bacteriophage ϕX174 origin region, nucleotides 4299–4328 of the ϕX174 DNA sequence, with its different functional domains. The *arrow* indicates the gene *A* protein cleavage site. The complete 30-bp origin is both sufficient and required for in vivo rolling circle DNA replication and DNA packaging. Termination takes place on the first 24 bp of the origin region

this region nucleotide changes can occur indicates that not all 30 nucleotides are essential for replication. Several different approaches have been used to study the nucleotide requirements for origin function. It should be noted that gene *A* protein is not only involved during initiation, i.e., cleavage of the viral strand at the origin, but also during elongation and termination of viral strand synthesis (IKEDA et al. 1976). Together with the *rep* protein, gene *A* protein participates in unwinding of the DNA helix ahead of the replication fork. In the termination reaction gene *A* protein covalently bound at the tail of the rolling circle interacts with the regenerated origin of the displaced single-stranded DNA chain. In a cleavage and ligation reaction circular viral strands are formed, whereas gene *A* protein is transferred to the newly synthesized viral strand (EISENBERG et al. 1978). So in some of the approaches and ways of analysis, not only initiation but also elongation, termination, and even packaging are measured.

The boundaries of the origin region have been determined in our group and in HURWITZ's laboratory. We have used synthetic oligonucleotides to construct plasmid DNAs containing either the entire 30-bp origin region or parts of it. These plasmid DNAs were tested *in vitro* as substrates for the ϕX174 gene *A* protein. These studies showed that the presence of the first 27 nucleotides of the origin region in a superhelical DNA molecule is both sufficient and required for ϕX174 gene *A* protein cleavage (HEIDEKAMP et al. 1981; FLUIT et al. 1984). Plasmid DNAs containing the first 26 nucleotides and shorter partial origin regions were not cleaved by gene *A* protein. In HURWITZ's laboratory recombinant DNA plasmids containing the isometric viral-strand origin region were constructed using DNA restriction fragments of ϕX174 and G4 (ZIPURSKY et al. 1980; BROWN et al. 1982, 1983). In addition to the ϕX174 gene *A* protein cleavage reaction these plasmids have also been tested as templates in the RF→ SS(c) *in vitro* DNA replication system. Plasmid DNA carrying the first 28 nucleotides of the origin region is capable of supporting viral-strand synthesis at levels similar to that of the ϕX174 RF DNA control. Plasmids containing the first 27 nucleotides of the origin region have not been tested in this system. A plasmid missing the first cytosine residue of the origin region was also isolated. This plasmid is cleaved by ϕX174 gene *A* protein in a similar way as the ϕX174 RF DNA control, but strongly reduced RF→SS(c) synthesis is observed due to poor reinitiation. These results clearly show that the first 28 nucleotides (and probably also the first 27) of the origin region are, at least *in vitro,* sufficient and necessary for ϕX gene A protein-mediated initiation, elongation, termination, and reinitiation of rolling circle DNA replication.

During stage III replication viral strands are replicated directly into phage proheads. FLUIT et al. (1985) have recently shown that for *in vivo* stage III replication the presence of the complete 30-bp origin is required. This is evaluated by measuring the transducing particles present in a ϕX174 or G4 lysate, obtained after infection of *E. coli* cells containing plasmids with complete and partial origins. A 20- and 200-fold reduction of the transduction frequency in the ϕX174 and G4 systems, respectively, is seen for a plasmid containing the first 28 nucleotides of the origin region compared to a plasmid with the whole 30-bp origin region. These results suggest that the 3'-end of the origin has an important function during stage III replication. It may be involved as a morphogenetic signal in the attachment of the prohead to the ϕX174 RFII-gene A complex.

The experiments described above already demonstrate that within the origin region different functional domains can be discerned. The dissection of the origin region into different domains was initiated by the observation that gene A protein cleaves a synthetic oligonucleotide containing the first 10 nucleotides of the origin region (VAN MANSFELD et al. 1980). This showed that gene A protein is a single-strand-specific endonuclease. Double-stranded DNA is only cleaved when the DNA contains a superhelical configuration, and this cleavage requires the presence of the 27 nucleotides of the origin region (see above). Further studies about the nucleotide sequence requirements for gene A protein cleavage of single-stranded DNA have led to the following consensus sequence:

$$
\begin{array}{ccccc}
\text{A} & & \text{T} \downarrow & \text{T} & \\
& \text{ACT} & \text{G} & \text{A} & \\
\text{T} & & \text{C} & \text{G} &
\end{array}
$$

for recognition by the protein (VAN MANSFELD et al. 1984a). The first cytosine residue of the conserved origin region is not included in the recognition sequence of gene A protein. This is in agreement with the observation that plasmid DNA containing 2–30 bp of the origin region is cleaved by ϕX174 gene A protein (see above). The recognition sequence of ϕX174 gene A protein is degenerate, which is also shown by the existence of a viable ϕX174 mutant with a T→C base change in position 6 of the origin region (BAAS et al. 1981a) (Table 2). The nucleotide specificity for cleavage of single-stranded DNA by the gene A protein is dramatically changed by the addition of SSB. In the presence of SSB, cleavage of single-stranded DNA requires, as in the case of double-stranded DNA, the presence of the first 27 nucleotides of the origin region (VAN MANSFELD et al. 1986a). Enhancement of the nucleotide specificity caused by SSB has also been observed in the initiation reaction of the complementary-strand synthesis of single-stranded DNA phages (KORNBERG 1980, 1982). Cleavage by gene A protein of the regenerated origin during termination of rolling circle DNA replication requires almost the complete origin sequence (see below). This may be explained by interference of SSB, which is bound to the displaced tail of the rolling circle, with the cleavage reaction.

Table 2. Bacteriophage ϕX174 ori-mutants constructed by oligonucleotide-directed mutagenesis

Mutant position	Nucleotide change	Amino acid change in gene A protein	Burst size
ori-6[a]	T → C	none	normal
ori-10	A → T	Ile → Phe	reduced
ori-11 (St-1, α₃)[b]	T → A	Ile → Asn	–
ori-12[c]	T → C	none	normal
ori-13	A → T	Asn → Tyr	normal
ori-14 (St-1, α₃)[b]	A → G	Asn → Ser	–
ori-14.1	A → T	Asn → Ile	normal
ori-14.2	A → C	Asn → Thr	normal
ori-14.3	A → G	Asn → Ser	normal
ori-15.1	T → C	none	reduced
ori-15.2	T → A	Asn → Lys	normal
ori-17	A → C	Asn → Thr	normal
	17A → T	Asn → Ile	
ori-17-19			reduced
	19A → G	Thr → Ala	
ori-18	C → T	none	reduced
ori-19.1	A → G	Thr → Ala	normal
ori-19.2	A → C	Thr → Pro	normal
ori-20.1	C → T	Thr → Ile	normal
ori-20.2	C → G	Thr → Ser	normal
ori-21	T → A	none	normal
ori-22	A → C	Ile → Leu	normal

No viable mutants could be isolated using oligonucleotide-directed mutagenesis with primers containing the following base changes:

7	G → A	Asp → Asn	
8	A → G or T	Asp → Gly or Val	
9	T → G	Asp → Glu	
15	T → G	Asn → Lys	
24	A → T	None	
27	C → T	None	

[a] Former notation m 402 (BAAS et al. 1981 a)
[b] Nucleotide changes found in the origin region of bacteriophages St-1 and α₃ are included in this table (HEIDEKAMP et al. 1980, 1982)
[c] Former notation m 316 (BAAS et al. 1981 a)

   The observation of the different sequence requirements for cleavage of double- and single-stranded DNA by gene A protein and the isolation of several viable ϕX174 mutants in the origin region (BAAS et al. 1981 a) (Table 2) have led to the following model for gene A protein interaction with the origin region during initiation of rolling circle DNA replication (BAAS et al. 1981 b; HEIDE-KAMP et al. 1982). The origin region can be divided into three domains: the recognition sequence of gene A protein (nucleotides 2–9), an AT-rich spacer region in which many nucleotide substitutions are tolerated (nucleotides 10–17), and a key binding sequence of gene A protein (nucleotides 18–27) (Fig. 5). During initiation of DNA replication gene A protein first binds to the binding sequence. This binding induces local denaturation, which is facilitated by the superhelical turns of the DNA and the AT-rich spacer region. Unwinding of this region exposes the recognition sequence in a single-stranded state. Gene *A* protein then cleaves the phosphodiester bond between the G and A residues in positions 7 and 8 of the origin region. After cleavage the G residue contains a free 3'-OH group which serves as a primer for DNA polymerase III holoenzyme, whereas the A residue is covalently bound to the gene *A* protein.

Two experimental data support the existence of a spacer sequence. Firstly, nucleotide changes at almost any position of this region have yielded viable phage mutants (BAAS 1987) (Table 2). For instance at position 14 the adenosine residue has been replaced by a cytosine, thymidine, or guanosine residue without great influence on the viability of the phage. Many of these mutants also change the amino acid sequence of the gene *A* protein, indicating that the corresponding region of the protein is not essential for the enzymatic reactions. Although many nucleotide substitutions in this region are allowed, the nature of the nucleotides is not completely unimportant for the viability of the phage. A T→C substitution in position 15 does not change the amino acid sequence, but results in a mutant phage (ori-15.1) with a reduced growth ability. A change of the same T residue into an A residue results in a mutant phage (ori-15.2) with an altered gene *A* protein (Asn→Lys), but this mutant has a normal burst size and growth rate. No viable phage with a G residue in this position, which also results in an Asn→Lys change in gene *A* protein, could be isolated. Secondly, a short insertion of seven nucleotides and a deletion of one nucleotide in the spacer region constructed in plasmid DNA containing the ϕX174 origin do not impair cleavage by gene *A* protein *in vitro*. However, these mutations inhibit replication and/or morphogenesis of phage particles as shown by the low yield of transducing phage particles obtained from ϕX174-infected cells harboring these plasmids (BAAS 1987). This is another demonstration that cleavage by gene *A* protein *in vitro* is a necessary but not a sufficient condition for replication *in vivo*.

The rightward boundary of the binding sequence at position 27 has been determined by analysis of the cleavage reaction of gene *A* protein on plasmid DNA containing different parts of the origin region. Plasmid DNA containing the first 26 nucleotides of the origin region is not cleaved by the gene *A* protein, whereas plasmid DNAs with 27, 28, or the complete conserved origin region are nicked (FLUIT et al. 1984). Using oligonucleotide-directed mutagenesis, it was also not possible to change the cytosine residue in this position into a thymidine residue (Table 2). Another line of evidence indicating the essential position of nucleotide 27 in the gene *A* protein-DNA interaction is the observation that pAF26 is not cleaved by gene *A* protein (FLUIT et al. 1984). pAF26 contains the first 26 nucleotides of the origin region followed by a GC sequence. Thus pAF26 can be considered to contain the first 28 bp of the ϕX origin region with a C→G substitution on position 27.

The leftward boundary of the binding sequence of gene *A* protein has been tentatively placed at position 18. A C→T substitution on position 18 has no effect on the amino acid sequence of gene *A* protein but results in a ϕX174 phage (ori-18) with a strongly reduced growth ability (10% of the burst size of the wild-type phage). Substitutions in the preceding and following nucleotides have no profound effect on the growth ability and burst size of the resulting ϕX174 mutant phages (Table 2). Recent mutational analysis of the binding sequence has shown that many base substitutions within this region are allowed. One interpretation of these results is that only a few of the contacts of gene A protein with the binding sequence require specific nucleotides (e.g., the nucleotides at positions 18, 24, and 27). Nucleotide changes at other positions also in the spacer region result in slightly reduced recognition and action of the gene A protein. For the majority of the isolated ori-mutants, no significant difference in growth rate and burst size compared to wild-type ϕX174 could be detected. However, in a Darwinian-type experiment in which wild-type ϕX174 together with several ori-mutants infect E. coli cells many times successively, the wild-type ϕX174 finally outgrows all the mutants (BAAS 1987). Also a ϕX174 ori-mutant (ori-17–19, Table 2) with two base changes in the origin region has a significantly lower burst size (50%) compared to the single mutants. These observations show the flexibility of the nucleotide sequence requirements for ϕX174 gene A protein action and on the other hand emphasize the superiority of the wild-type ori-sequence during reproduction of the isometric phages.

6.2 Nucleotide Sequence Requirements for Termination of Rolling Circle DNA Replication of the Isometric Phages

At the end of a replication cycle the bound gene A protein recognizes the regenerated origin sequence and terminates the replication by a cleavage and ligation reaction. The old viral strand is displaced as a circle, whereas the gene A protein is transferred to the newly synthesized viral strand. The nucleotide sequence requirements for termination of rolling circle DNA replication and transfer of gene A protein have been studied using partial origin sequences. BROWN et al. (1982, 1984) have observed, in the RF→SS(c) in vitro DNA replication system, gene A protein transfer to a partial origin sequence of the first 25 nucleotides. This region is constructed during DNA replication by incorporation of dideoxyguanosine triphosphate (ddGTP) at position 25 of the origin, which prevents further DNA replication. Unwinding of the DNA duplex, however, continues, and gene A protein transfer could be detected after nuclease treatment by the presence of gene A protein covalently bound to a short radioactive oligonucleotide. The old viral strand can be traced as circular, single-stranded DNA through hybridization studies. Incorporation of dideoxycytidine triphosphate (ddCTP) at position 18 completely abolishes gene A protein transfer in the same system. FLUIT et al. (1986) studied the termination reaction using plasmid DNA containing a complete and a partial origin sequence. Termination at the partial origin sequence was investigated by analyzing transducing particles obtained after infection of E. coli harboring such plasmids with bacteriophage ϕX174 or G4. Termination of rolling circle DNA replication occurs efficiently on those consisting of 24 nucleotides. A partial origin sequence consisting of the first 16 nucleotides is not recognized as a termination signal in agreement with earlier in vitro DNA replication studies (REINBERG et al. 1983). Termination at sequences which show some homology with the origin sequence is observed with low frequency for the *in vivo* (BAAS 1987) and the *in vitro* (AOYAMA and HAYASHI 1985) packaging systems.

6.3 Nature of the Covalent Linkage Between DNA and Gene *A* Protein in the Isometric Phages

The existence of a covalent linkage between gene *A* protein and the origin DNA is concluded from a number of different observations. It is impossible to use the 5′-end of the cleaved viral strand after treatment with alkaline phosphatase as a substrate for T4 polynucleotide kinase (LANGEVELD et al. 1978). Also the RFII-*A* complex is not sealed after incubation with T4 DNA ligase, and no reaction is observed with DNA polymerase I in the presence of the four deoxyribonucleoside triphosphates (dNTP's) (IKEDA et al. 1976, 1979). The linkage is not disrupted by treatment of the complex with 0.2 *M* NaOH, high salt solutions (CsCl equilibrium density gradient), or boiling in the presence of 1% sodium dodecyl sulfate (SDS) (EISENBERG and KORNBERG 1979). Radioactive measurements show that after cleavage only one gene *A* protein molecule is bound to the DNA (EISENBERG and KORNBERG 1979). Participation of multimeric forms of gene *A* protein during the cleavage reaction has been proposed (IKEDA et al. 1979; ROTH et al. 1984).

A* protein, a second product of gene *A* resulting from an internal in-frame translation start (LINNEY and HAYASHI 1973), retains a number of the activities of gene *A* protein. For a long time the function of A* protein has remained unclear. Recently, a specific role for A* protein in the switch from stage II to stage III ϕX174 DNA replication has been suggested (VAN MANSFELD et al.

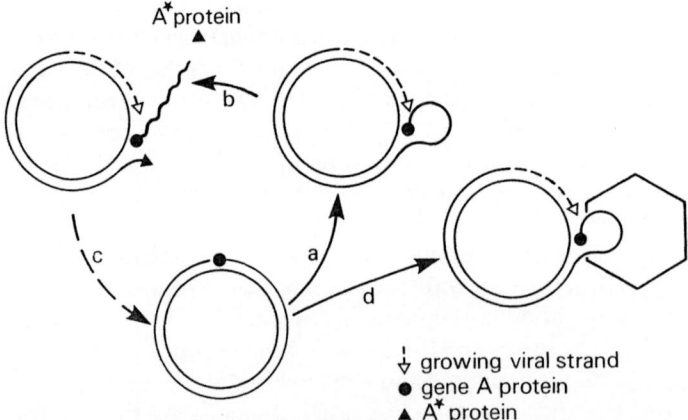

Fig. 6. Model for the function of A* protein during stage III ϕX174 DNA replication. Stage III involves rolling circle DNA replication coupled to packaging of single-stranded DNA into phage coats. No RF DNA replication takes place during this stage. The model predicts that RF DNA replication is interrupted by *A** protein cleavage at a preferential cleavage site in the intergenic region between genes *J* and *F*. This leads to futile replication cycles until sufficient coat proteins are available to package the single-stranded DNA into phage coats, which prevents *A** protein cleavage

1986a) (Fig. 6). A* protein cleaves ϕX174 DNA at the origin and at several
other sites (LANGEVELD et al. 1979, 1981). A* protein also binds covalently
to the 5'-end of the DNA at the cleavage site (VAN MANSFELD et al. 1980)
and has ligating activity (LANGEVELD et al. 1980; EISENBERG and FINER 1980;
VAN MANSFELD et al. 1982). The linkage between the DNA and the gene A
and A* proteins was identified as a phosphodiester bond between a tyrosine
residue of the protein and the 5'-phosphate of the adenosine residue at the
cleavage site (VAN MANSFELD et al. 1984b; ROTH et al. 1984; SANHUEZA and
EISENBERG 1984, 1985; ZOLOTUKHIN et al. 1984). The characterization of the
DNA-protein linkage was done independently in four laboratories using slightly
different approaches and substrates. In each approach the phosphodiester bond
between the G and A residues at the cleavage site in an oligonucleotide or
in ϕX174 DNA is made radioactive. After cleavage of the substrate with gene
A or A* protein, the radioactive protein-DNA complex is isolated. Digestion
with nucleases and proteases followed by complete acid hydrolysis in $6N$ HCl
leads to the characterization of phosphotyrosine as the major labeled product.
After digestion of the radioactive protein-DNA complex with trypsine or pro-

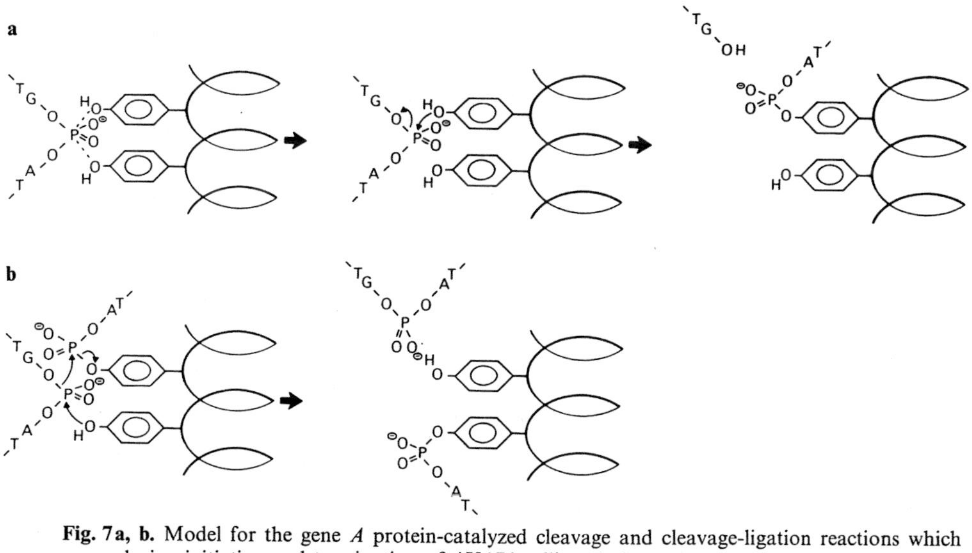

Fig. 7a, b. Model for the gene A protein-catalyzed cleavage and cleavage-ligation reactions which
occur during initiation and termination of ϕX174 rolling circle replication, respectively. The *origin
sequence* is shown schematically as: $-TG-O-P-O-AT-$.

The two tyrosyl residues at the active center of the gene A protein are supposed to be part of
an α-helix which brings their side-chains into juxtaposition. **a** Cleavage. Binding of gene A protein
with DNA places the phosphorus atom at the origin in a position symmetrical towards the two
tyrosyl-OH groups. Nucleophilic attack (*arrow*) leads to transesterification as shown. The results
indicate that either one of the oxygen atoms of the two tyrosyl residues can function as the acceptor
of the DNA chain in this reaction. **b** Cleavage-ligation. After one round of replication gene A
protein binds the regenerated origin sequence. A nucleophilic attack (*arrow*) of the hydroxyl group
of the free tyrosyl residue initiates two successive transesterifications as indicated. This results in
a phosphodiester bond between DNA and this tyrosyl residue, a free hydroxyl group at the other
tyrosyl residue and the formation of a phosphodiester bond which ligates the two DNA ends

teinase K, two ^{32}P-labeled peptides are found in approximately equimolar ratios. This observation strongly suggests that two different tyrosine residues of the gene *A* protein are involved in the cleavage reaction.

VAN MANSFELD et al. (1986b) recently determined the amino acid sequence of the peptide moieties of the two peptide-oligonucleotide complexes. Inspection of the amino acid sequence of gene *A* protein shows that the two peptides met, ala, val, gly, phe, tyr, val, ala, lys and tyr, val, asn, lys are located adjacent to each other and correspond to amino acid residues 338–350. The two tyrosine residues lie only three amino acid residues apart in the repeating sequence: tyr, val, ala, lys, tyr, val, asn, lys. If this part of the gene *A* protein occurs in an α-helix conformation, both tyrosyl side-chains protrude from the same side of the molecule. Figure 7 shows a model for the successive gene *A* protein-catalyzed cleavage and cleavage-ligation reactions which take place during initiation and termination of rolling circle DNA replication. The model shows how the two juxtaposed tyrosyl residues play an equivalent role, and how the two tyrosyl-hydroxyl groups cooperate as two active groups in one active center.

6.4 The Ff Phage Viral-Strand Origin

Comparison of the nucleotide sequences in the region around the gene II protein cleavage site in bacteriophages f1, fd, and M13 does not provide a clue to the limits of the viral-strand origin (HORIUCHI et al. 1979; SCHALLER 1979; SUGGS and RAY 1979). In contrast to the isometric phages the nucleotide sequence of the filamentous Ff phages is strongly conserved. The boundaries of the viral-strand origin are determined by deletion analysis of the replication origin cloned in pBR322 DNA. The presence of a functional viral-strand origin on a plasmid interferes with the replication of superinfecting phage. This interference, caused by competition for the host and phage replication proteins, results in a reduced phage yield, production of transducing phage particles, and stimulation of plasmid DNA synthesis. Plasmid DNA enters the filamentous mode of replication, which gives rise to more plasmid DNA molecules within the infected cell and ultimately to transducing particles. Another way to measure the origin function is by the ability of the ori-plasmid to transform a filamentous phage-infected polA$^-$ strain. This situation is nonpermissive for the pBR322 replicon. Transformation can only occur under the control of the cloned filamentous phage sequences on the pBR322 derivative (CLEARY and RAY 1980). The 5′- and 3′-deletion studies of the *Hpa*II H DNA restriction fragment have located the functional filamentous phage origin in a DNA fragment of approximately 140 bp (CLEARY and RAY 1981; KIM et al. 1981; DOTTO et al. 1982a, 1984). Its 5′-side is located 12 nucleotides upstream from the gene II protein cleavage site. The effect of various deletions and insertions within the 140-bp region was also studied (DOTTO et al. 1984; JOHNSTON and RAY 1984). Besides the above-mentioned tests the mutated origins are also used as substrates for the purified gene II protein *in vitro*. In the presence of Mg^{2+}, RFI DNA is converted by gene II protein to equal amounts of RFII and RFIV DNA (RF DNA with both strands closed, containing no superhelical turns) (MEYER and GEIDER

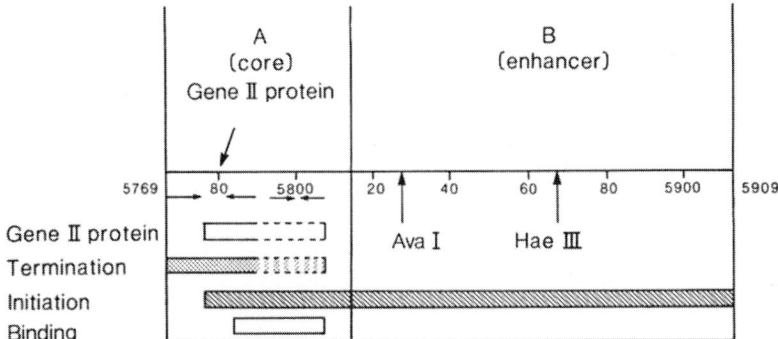

Fig. 8. The Ff functional origin region. *Numbers* indicate nucleotide positions on the f1, M 13 map. The *downwards arrow* indicates the gene II protein cleavage site. The location of the gene II protein recognition sequence, the gene II protein-binding sequence, and the location of the signals for initiation and termination of viral-strand synthesis are indicated. Also shown are the *Ava*I and *Hae*III restriction sites used for insertion and deletion mutagenesis. (Adapted from Dotto et al. 1982a; Horiuchi 1986)

1979a, b). Besides the cleavage and nicking-closing reaction, a binding assay for gene II protein was also developed (Horiuchi 1986). Defects in initiation or termination of the mutated origins could be studied by constructing chimeras with a wild-type and a mutated origin in the same orientation (Dotto et al. 1982b). Infection of a bacterium containing a chimera with two wild-type origins results in the segregation of this plasmid into two replicons (Dotto and Horiuchi 1981). Rolling circle DNA replication initiated at the first origin will terminate at the second origin, leading to breakdown of the original chimera. Analysis of the length of the plasmid DNA observed after infection of a bacterium harboring chimeras with wild-type and mutated origins made it possible to decide whether initiation or termination or both are affected by the mutation (Dotto et al. 1982b, 1984). Examination of the data obtained with many mutants leads to a dissection of the origin region into two different functional domains (Fig. 8): domain A, the core sequence of approximately 45 nucleotides, and domain B, the enhancer sequence of approximately 100 nucleotides. Domain B is not absolutely required but rather facilitates, by a still unknown mechanism, the initiation of viral-strand replication. *In vitro*, DNAs from mutants in domain B bind and are cleaved by gene II protein even at limiting concentrations. Superinfection of cells harboring chimeras with a mutated domain B results in a normal yield of phage particles of which 1% are transducing particles. Superinfection of cells harboring chimeras with a mutated domain A also produces normal yields of phage particles, but the amount of transducing particles is lower than 0.1%. The 45 nucleotides of domain A can be drawn in two hairpin structures (Fig. 9); in the literature these hairpins are denoted D and E or A1 and A2. The complete intergenic region of the filamentous phages can be drawn as five hairpin structures (A, B, C, D, and E in the order of the 5′ to 3′ direction of the plus strand). Hairpin A contains the morphogenetic signal (Dotto et al. 1981; Dotto and Zinder 1983) and a *rho*-dependent tran-

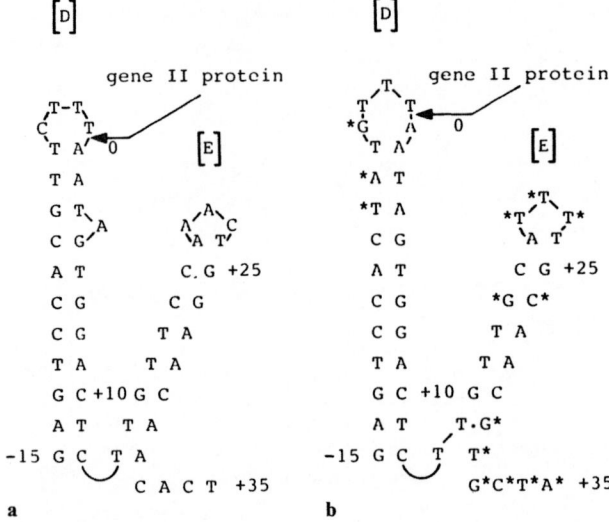

Fig. 9a, b. Possible secondary structure of DNA sequences at the viral-strand origin (domain A or core) of the filamentous Ff phages (**a**) and IKe (**b**). The *arrow* indicates the gene II protein cleavage site (position *zero*). The *asterisks* in the IKe DNA sequence denote nucleotide differences from the Ff phages

scription termination signal (MOSES and MODEL 1984; SMITS et al. 1984) and hairpins B and C constitute the complementary-strand origin (see above). The top of the palindrome D contains the gene II protein cleavage site, which is used as reference point zero. Domain A can be subdivided into four distinct, but partially overlapping sequences (Fig. 8): a sequence required for binding of gene II protein, a sequence required for *in vitro* cleavage by gene II protein, a sequence required for initiation of viral-strand synthesis, and a sequence required for termination of viral-strand synthesis.

Although the exact boundaries of the different sequences have not been established, the data obtained with various deletion mutants clearly show different nucleotide sequence requirements for the above-mentioned aspects in the process of DNA replication. Termination of viral-strand synthesis requires nucleotides -12 to $+(11-29)$. For the termination reaction the presence of the complete palindrome D is necessary (DOTTO et al. 1982b, 1984). This palindromic sequence may be necessary to obtain a proper conformation of the DNA at the end of a replication cycle for gene II protein cleavage. It may be similar to the structure of partially melted, double-stranded DNA in a superhelical molecule. In contrast to the gene *A* protein of the isometric phages, gene II protein does not cleave single-stranded DNA (Table 1). Alternatively, this sequence may be required after cleavage to bring the 5'- and 3'-ends of the single-stranded DNA molecule together for circularization. The *in vitro* gene II protein cleavage sequence extends from nucleotide -3 to nucleotide $+(11-29)$ (DOTTO et al. 1982a; PEETERS et al. 1987). The cleavage sequence contains half of palindrome D and part of or almost the complete palindrome

E. Note that the 3′-end of the termination sequence signal and the gene II protein cleavage sequence coincide. A binding sequence for gene II protein within domain A is localized by trapping DNA-gene II protein complexes on nitrocellulose filters (HORIUCHI 1986). Surprisingly, the binding does not require superhelical DNA. The sequence responsible for the binding could be found between nucleotides $+5$ and $+29$, involving the region around palindrome E, possibly including the adjacent quarter of palindrome D. The binding sequence does not need the sequence at the cleavage site. The right end of the binding sequence may be identical to that of the gene II protein cleavage sequence. A correlation between the *in vitro* binding studies and the following *in vivo* data is observed. Defective deletion ori-mutants containing an intact gene II protein-binding sequence inhibit growth of superinfecting phage, if the intracellular concentration of gene II protein is low (HORIUCHI 1986). In addition these deletion mutants, inserted into a plasmid that already has a functional filamentous origin, prevent rolling circle DNA replication of the plasmid upon superinfection. These *in vivo* observations can be explained by binding of the gene II protein to the defective origins. Owing to the binding of gene II protein at the defective origin on the plasmid, virtually no gene II protein molecules are left for the replication of the superinfecting phage DNA, and this binding may block DNA replication of the plasmid initiated from the wild-type origin.

The sequence required for initiation of viral-strand synthesis extends from -3 to $+(29-40)$. The 5′ boundary coincides with the gene II protein cleavage sequence, and the 3′ boundary may extend a few nucleotides beyond the palindrome E. This region is presumably involved in the formation or initial movement of the replication fork. In vitro synthesis of viral strands after cleavage at the origin still requires the presence of gene II protein (MEYER and GEIDER 1982). This sequence may be necessary for the interaction of gene II protein with the *rep*, DNA polymerase III holoenzyme, and SSB protein.

6.5 Secondary Mutations Reduce the Length of Ff Viral-Strand Origin Sequence to Domain A

As described above domain B is not absolutely required for viral-strand synthesis. Disruption of this domain on an ori-plasmid by insertions or deletions at the *Ava*I restriction site at position $+45$ or the *Hae*III restriction site at position $+88$ reduces the biological activity of the origin to the level of domain A alone (DOTTO et al. 1984; JOHNSTON and RAY 1984). Domain B facilitates viral-strand synthesis as an "enhancer." The mechanism which is responsible for this activation is not understood but certainly involves gene II protein. Filamentous phages have been constructed and isolated which bypass the need for domain B for their replication. These phages require only the core of the functional origin, while domain B is completely dispensable (DOTTO and ZINDER 1984a, b). Some of these phages are spontaneous mutants which arise after cloning of exogenous DNA sequences in domain B (BOEKE et al. 1979; MESSING et al. 1977; MESSING and VIEIRA 1982). KIM and RAY (1985) selected for these phages by transformation of a phage-infected polA⁻ strain by ori-plasmids containing deletions in domain B. Analysis of them shows that the dependence of domain B has been lost by compensating mutations elsewhere in the phage genome. These mutations result in a change in the amino acid sequence of gene II protein or in a five- to tenfold higher concentration of gene II protein in the infected cell (DOTTO and ZINDER 1984a, b; KIM and RAY 1985). The increased concentration of gene II protein is either the result of a mutation in the gene V protein or a mutation in the gene II mRNA leader. Both mutations interfere with the repression of translation of the gene II mRNA. In wild-type infections gene V protein represses late in infection translation of the gene II mRNA by binding at the mRNA leader sequence (FULFORD and MODEL 1984; MODEL et al. 1982; YEN and WEBSTER 1982).

At this moment it is not clear why increased levels or an altered gene II protein inside the cell has such a drastic effect on the length of the functional origin. The sequences responsible for binding and cleavage by gene II protein are located within domain A. Domain A also contains a sequence at its 3′-end which is required for initiation of viral-strand synthesis. This sequence is probably involved in the formation or initial movement of the replication fork. The apparent role of domain B in this event is as yet obscure. Possibly, gene II protein interacts directly with nucleotide sequences in domain B, and an increased concentration of gene II protein or a mutated gene II protein might render such nucleotide sequences unnecessary by altering the DNA-protein interactions involved in the initiation process.

6.6 The IKe Phage Viral-Strand Origin

Recently the nucleotide sequence of the N-specific filamentous phage IKe has been determined (PEETERS et al. 1985). IKe displays the same genetic organization as the Ff phages. However, IKe is only distantly related to the Ff phages, as shown by the overall nucleotide sequence homology of 55%. Inspection of the nucleotide sequence of the viral- (Fig. 9B) and complementary-strand origin (Fig. 4B) of IKe shows a strong conservation with respect to the origins of the Ff phages. With a few variations domain A of the viral-strand origin is conserved in bacteriophage IKe; a nucleotide sequence homologous to domain B is not present. Deletion studies have shown that a plasmid containing a sequence of 49 nucleotides downstream from the gene II protein cleavage site has retained full origin activity (PEETERS et al. 1986a). This strongly suggests that for IKe DNA replication, no enhancer sequence is required. In spite of the high degree of homology between the gene II proteins of IKe and Ff (approximately 60%) and the conservation of domain A of the origin region, the biological functions of these proteins are not interchangeable (PEETERS et al. 1986a). Gene II protein of bacteriophage IKe, provided by a recombinant plasmid, does not complement Ff am mutants in gene II and vice versa. Also, Ff ori-plasmids do not replicate in polA⁻ cells infected with bacteriophage IKe. These observations have been used to design a cloning vector pKUN9 containing the viral-strand origin of Ff and IKe and their respective morphogenetic signals in an opposite orientation (PEETERS et al. 1986b). Infection of cells harboring these plasmids with IKe or Ff results in the selective packaging of either DNA strand in phage-like particles.

Although the gene II protein of Ff phage cannot replicate the viral strand of IKe, it retains the ability to cleave *in vitro* specifically superhelical RFI IKe DNA at the origin. *In vivo* studies have shown that Ff gene II protein also recognizes the IKe origin during termination of replication and vice versa (PEETERS et al. 1987). Recombinant plasmids were constructed containing the viral-strand replication origin of IKe and Ff in the same orientation. Infection of cells harboring these plasmids with Ff phage will lead to rolling circle replication of these plasmids. Replication is initiated at the Ff origin and terminates at the IKe origin, as shown by the appearance of plasmids with fusion origins. The DNA of this fusion origin upstream from the gene II cleavage site is derived from IKe. The DNA at the 3'-side of the gene II protein cleavage site is obtained from the Ff phage. The nucleotide sequence downstream from the gene II protein cleavage site of these fusion origins determines the specificity of the ori-function with respect to the IKe or Ff phage. The first nucleotide differences between Ff and IKe are found in palindrome E (Fig. 9). The nucleotide change at position +18 does not disturb the potential to form a secondary structure because of the compensating mutation at position +26. This strongly suggests the importance of the secondary structure for origin function. DOTTO et al. (1982a, 1984) have shown that a plasmid pΔ +29, containing an origin sequence up to position +29, is cleaved by gene II protein but is unable to initiate viral-strand synthesis. Therefore, DNA sequences downstream from position +29 are essential to initiate viral-strand synthesis. A number of nucleotide differences exist there between the IKe and Ff replication origins. This strongly suggests that this region is also important for the determination of the specificity of the origin region. A contribution to the specificity of the nonconserved nucleotides in the stem and loop of the hairpin structure E, however, cannot be excluded. Site-directed mutagenesis of these nonconserved nucleotides should answer this question.

7 Concluding Remarks

The genomes of the single-stranded DNA phages replicate in an autonomous fashion, i.e., independently of the chromosomes of their host cells. The phage DNA contains the two essential elements for autonomous replication as predicted by the replicon model (JACOB et al. 1963): firstly, an initiator gene coding for the initiator protein and, secondly, an origin which is recognized by the initiator protein for the start of replication.

With the present detailed knowledge of the replication cycle of single-stranded DNA phages, one can ask the question whether or not the initiator gene and origin region are the only loci of the phage genomes which are required for autonomous DNA replication and its regulation. Regulation takes place by controlling the number of replicating intermediates and the different modes of DNA synthesis in stages I, II, and III of the life cycle of single-stranded DNA phages.

For the isometric phages it has been suggested that stages I and II replication require binding of the phage DNA to an essential membrane site. There are supposed to be a limited number of these sites which may explain the fact that only a few, presumably one, (parental) RF DNA molecule per cell can replicate during stage II replication. The view that particular "replication sites" within the cell are required for isometric phage DNA replication finds strong support in the discovery of the so-called "reduction sequence" corresponding to the intergenic region between genes *H* and *A* of the ϕX174 and G4 phage genomes, respectively (VAN DER AVOORT et al. 1982, 1984). These authors have shown that in cells containing plasmids with this reduction sequence, stage I replication of the corresponding phage is strongly inhibited. The data suggest that this inhibition is due to the blocking by the plasmid DNA of cellular sites which are required for the replication of the phage DNA and that there are only a limited number of these "replication sites" per cell. So, the "reduction sequence" may be another phage locus involved in the regulation of phage DNA replication.

Studies with *in vitro* systems have not revealed any evidence for the requirement of the "reduction sequence" in stages I and II replication, indicating that *in vitro* studies may not disclose all the complexities of the situation *in vivo*. Similarly, *in vitro* systems for stage II (BROWN et al. 1983) and stage III replication (AOYAMA et al. 1983) have shown that only part of the 30-bp origin region of the isometric phage DNA is required. However, for efficient stage III replication *in vivo* the complete 30-bp origin region is required (FLUIT et al. 1985), again indicating that signals of the phage genome which play a role *in vivo* may remain unnoticed by the *in vitro* systems.

Finally, the A* protein cleavage site which is located between nucleotides 990 and 991 of the ϕX174 DNA sequence should be mentioned. It has been shown that the A* protein cleaves ϕX174 single-stranded DNA at this site late in the infection cycle (VAN MANSFELD et al. 1982, 1986a). The experimental evidence suggests that the A* protein plays a specific role in the switch from stage II to stage III ϕX174 DNA replication (Fig. 6). Therefore, the A* protein recognition sequence may be another locus in the ϕX174 DNA sequence which is required for the regulated autonomous replication of the single-stranded DNA phages.

These examples for the isometric phages show that our knowledge concerning the regulation of phage DNA replication is far from complete and that even these relatively simple phage DNAs may offer many avenues for fruitful further research. Also, as stated by ARTHUR KORNBERG: "Perhaps many of the still missing parts of the replication puzzle in both animal and bacterial cells will be seen through windows opened by research with viruses" (KORNBERG 1980).

Note Added in Proof

Site directed mutagenesis of the ATG start codon of A* protein of bacteriophage ϕX174 has shown, that A* protein is not essential for ϕX174 reproduction (BAAS et al. 1987 FEBS Lett 218:119–125). A viable phage mutant ϕX-4499T containing a G→T substitution at position 4499 of the ϕX174 DNA sequence was obtained which does not synthesize A* protein. However, the burst size of this mutant amounts to 50% of the burst size of wild type ϕX174, indicating the profitable role of A* protein in wild type infection. A* protein inhibits *E. coli* DNA replication thereby mobilizing the host replication factors for phage DNA replication and A* protein may be involved in an efficient switch from stage II into stage III replication. Colasanti and Denhardt came to the same conclusion by alteration of the ATG startcodon of A* protein into a TAG stopcodon (COLASANTI J, DENHARDT DT (1987) J. Mol. Biol. 197:47–54). The comparable gene X protein of the filamentous phages, however, is essential for viral strand replication late in infection (FULFORD W, MODEL P (1984) J. Mol. Biol. 178:137–153).

Aoyama and Hayashi showed that ϕX174 gene C protein inhibits *in vitro* viral strand DNA synthesis by binding after a replication round to the replication complex. Addition of proheads to the RF→SS(c) in vitro replication system results in a start of stage III replication. So the amount of ϕX174 gene C protein seems to be an important factor in the transition of stage II into stage III replication (AOYAMA A, HAYASHI M (1986) Cell 47:99–106).

The sequence TGGACTCTTGTTCCA, $+6$ to $+20$ of the Ff viral-strand origin, cloned in a plasmid is sufficient to bind presumably 2 molecules gene II protein, as shown by gel retardation and DNaseI footprint analysis. A similar DNaseI footprint with the complete Ff viral-strand origin was obtained with low concentration of gene II protein (complex I). At higher concentration of gene II protein complex II is formed containing presumably 4 molecules gene II protein. Complex II protects about 40 basepairs (from -7 to $+33$) from DNaseI digestion. Formation of this complex requires a DNA sequence which extends in the 3′ direction a few nucleotides beyond position $+29$ (GREENSTEIN D, HORIUCHI K (1987), J. Mol. Biol. 97:157–174).

References

Abarzua P, Marians KJ (1984) Enzymatic techniques for the isolation of random single base substitutions *in vitro* at high frequency. Proc Natl Acad Sci USA 81:2030–2034

Abarzua P, Soeller W, Marians KJ (1984) Mutational analysis of primosome assembly sites: I. Distinct classes of mutants in the *Escherichia coli* factor Y DNA effector sequences. J Biol Chem 259:14286–14292

Allison DP, Ganesan AJ, Olson AC, Snyder CM, Mitra S (1977) Electron microscopic studies of bacteriophage M13 DNA replication. J Virol 24:673–684

Aoyama A, Hayashi M (1985) Effects of genome size on bacteriophage ϕX174 DNA packaging *in vitro*. J Biol Chem 260:11033–11038

Aoyama A, Hamatake RK, Hayashi M (1983) *In vitro* synthesis of bacteriophage ϕX174 by purified components. Proc Natl Acad Sci USA 80:4195–4199

Arai K, Kornberg A (1981) Unique primed start of ϕX174 DNA replication and mobility of the primosome in a direction opposite chain synthesis. Proc Natl Acad Sci USA 78:69–73

Arai K, Low RL, Kornberg A (1981) Movement and site selection for priming by the primosome in phage ϕX174 DNA replication. Proc Natl Acad Sci USA 78:707–711

Baas PD (1985) DNA replication of single-stranded *Escherichia coli* DNA phages. Biochim Biophys Acta 825:111–139

Baas PD (1987) Mutational analysis of the bacteriophage ϕX174 replication origin. J Mol Biol

Baas PD, Jansz HS (1972) ϕX174 replicative form DNA replication, origin and direction. J Mol Biol 63:569–576

Baas PD, Jansz HS, Sinsheimer RL (1976) Bacteriophage ϕX174 DNA synthesis in a replication-deficient host. Determination of the origin of ϕX DNA replication. J Mol Biol 102:633–656

Baas PD, Teertstra WR, Jansz HS (1978) Bacteriophage ϕX174 RF DNA replication *in vivo*: a biochemical study. J Mol Biol 125:167–185

Baas PD, Teertstra WR, Van Mansfeld ADM, Jansz HS, Van der Marel GA, Veeneman GH, Van Boom JH (1981a) Construction of viable and lethal mutations in the origin of bacteriophage ϕX174 using synthetic oligodeoxyribonucleotides. J Mol Biol 152:615–639

Baas PD, Heidekamp F, Van Mansfeld ADM, Jansz HS, Van der Marel GA, Veeneman GH, Van Boom JH (1981b) Essential features of the origin of bacteriophage ϕX174 RF DNA replication. In: Ray DS, Fox CF (eds) The initiation of DNA replication. Academic, New York, pp 195–209

Beck E, Zink B (1981) Nucleotide sequence and genome organization of filamentous bacteriophage fl and fd. Gene 16:35–58

Beck E, Sommer R, Auerswald EA, Kurz C, Zink B, Osterburg G, Schaller H, Sugino K, Sugisaki H, Okamoto T, Takanami M (1978) Nucleotide sequence of bacteriophage fd DNA. Nucleic Acids Res 5:4495–4503

Benbow RM, Zuccarelli AJ, Sinsheimer RL (1974) A role for single-strand breaks in bacteriophage ϕX174 genetic recombination. J Mol Biol 88:629–651

Benz EW Jr, Reinberg D, Vicuna R, Hurwitz J (1980a) Initiation of DNA replication by the *dna*G protein. J Biol Chem 255:1096–1106

Benz EW Jr, Sims J, Dressler D, Hurwitz J (1980b) Tertiary structure is involved in the initiation of DNA synthesis by the *dna*G protein. In: Alberts B, Fox CF (eds) Mechanistic studies of DNA replication and genetic recombination. Academic, New York, pp 279–291

Boeke JD, Vovis GF, Zinder ND (1979) Insertion mutant of bacteriophage fl sensitive to *Eco* RI. Proc Natl Acad Sci USA 76:2699–2702

Bouché JP, Zechel K, Kornberg A (1975) *dna*G gene product, a rifampicin-resistant RNA polymerase initiates the conversion of a single-stranded coliphage DNA to its duplex replicative form. J Biol Chem 250:5995–6001

Bouché JP, Rowen L, Kornberg A (1978) The RNA primer synthesized by primase to initiate phage G4 DNA replication. J Biol Chem 253:765–769

Bowman K, Ray DS (1975) Degradation of the viral strand of ϕX174 parental replicative form DNA in a *rep*⁻ host. J Virol 16:838–843

Brown DR, Reinberg D, Schmidt-Glenewinkel T, Roth MJ, Zipursky SL, Hurwitz J (1982) DNA structures required for ϕX174 A protein-directed initiation and termination of DNA replication. Cold Spring Harbor Symp Quant Biol 47:701–715

Brown DR, Schmidt-Glenewinkel T, Reinberg D, Hurwitz J (1983) DNA sequences which support activities of the bacteriophage ϕX174 gene A protein. J Biol Chem 258:8402–8412

Brown DR, Roth MJ, Reinberg D, Hurwitz J (1984) Analysis of bacteriophage ϕX174 gene A protein-mediated termination and reinitiation of ϕX174 DNA synthesis: I. Characterization of the termination and reinitiation reactions. J Biol Chem 259:10545–10555

Brutlag D, Schekman R, Kornberg A (1971) A possible role of RNA polymerase in the initiation of M13 DNA synthesis. Proc Natl Acad Sci USA 68:2826–2829

Cleary JM, Ray DS (1980) Replication of the plasmid pBR322 under the control of a cloned replication origin from the single-stranded DNA phage M13. Proc Natl Acad Sci USA 77:4638–4642

Cleary JM, Ray DS (1981) Deletion analysis of the cloned replication origin region from bacteriophage M13. J Virol 40:197–203

Danna KJ, Nathans D (1972) Bidirectional replication of simian virus 40 DNA. Proc Natl Acad Sci USA 69:3097–3100

Denhardt DT, Dressler D, Ray DS (1978) The single-stranded DNA phages. Cold Spring Harbor Laboratory, Cold Spring Harbor, New York

Dintzis HM (1961) Assembly of the peptide chains of hemoglobin. Proc Natl Acad Sci USA 47:247–261

Dotto GP, Horiuchi K (1981) Replication of a plasmid containing two origins of bacteriophage f1. J Mol Biol 153:169–176

Dotto GP, Zinder ND (1983) The morphogenetic signal of bacteriophage f1. Virology 130:252–256

Dotto GP, Zinder ND (1984a) Increased intracellular concentration of an initiator protein markedly reduces the minimal sequence required for initiation of DNA synthesis. Proc Natl Acad Sci USA: 81:1336–1340

Dotto GP, Zinder ND (1984b) The minimal sequence for initiation of DNA synthesis can be reduced by qualitative or quantitative changes of an initiator protein. Nature 311:279–280

Dotto GP, Enea V, Zinder ND (1981) Functional analysis of bacteriophage f1 intergenic region. Virology 114:463–473

Dotto GP, Horiuchi K, Jakes KS, Zinder ND (1982a) Replication origin of bacteriophage f1. Two signals required for its function. J Mol Biol 162:335–343

Dotto GP, Horiuchi K, Zinder ND (1982b) Initiation and termination of phage f1 plus-strand synthesis. Proc Natl Acad Sci USA 79:7122–7126

Dotto GP, Horiuchi K, Zinder ND (1984) The functional origin of bacteriophage f1 DNA replication. Its signals and domains. J Mol Biol 172:507–521

Duguet M, Yarranton G, Gefter M (1979) The *rep* protein of *Escherichia coli:* interaction with DNA and other proteins. Cold Spring Harbor Symp Quant Biol 43:335–343

Eisenberg S, Denhardt DT (1974a) Structure of nascent ϕX174 replicative form; evidence for discontinuous DNA replication. Proc Natl Acad Sci USA 71:984–988

Eisenberg S, Denhardt DT (1974b) The mechanism of replication of ϕX174 single-stranded DNA: X. Distribution of the gaps in nascent RF DNA. Biochim Biophys Res Commun 61:532–537

Eisenberg S, Finer M (1980) Cleavage and circularization of single-stranded DNA: a novel enzymatic activity of ϕX174 A* protein. Nucleic Acids Res 8:5305–5315

Eisenberg S, Kornberg A (1979) Purification and characterization of ϕX174 gene A protein. A multifunctional enzyme of duplex DNA replication. J Biol Chem 254:5328–5332

Eisenberg S, Harbers B, Hours C, Denhardt DT (1975) The mechanism of replication of ϕX174: XII. Non-random locations of gaps in nascent ϕX174 RFII DNA. J Mol Biol 99:107–123

Eisenberg S, Griffith J, Kornberg A (1977) ϕX174 cistron A protein is a multifunctional enzyme in DNA replication. Proc Natl Acad Sci USA 74:3198–3202

Eisenberg S, Scott JF, Kornberg A (1978) An enzyme system for replicating the duplex replicative form of ϕX174 DNA. In: Denhardt DT, Dressler D, Ray DS (eds) The single-stranded DNA phages. Cold Spring Harbor Laboratory, Cold Spring Harbor, New York, pp 287–302

Fiddes JC, Barrell BG, Godson GN (1978) Nucleotide sequences of the separate origins of synthesis of bacteriophage G4 viral and complementary strands. Proc Natl Acad Sci USA 75:1081–1085

Fluit AC, Baas PD, Van Boom JH, Veeneman GH, Jansz HS (1984) Gene A protein cleavage of recombinant plasmids containing the ϕX174 replication origin. Nucleic Acids Res 12:6443–6454

Fluit AC, Baas PD, Jansz HS (1985) The complete 30-base pair origin region of bacteriophage ϕX174 in a plasmid is both required and sufficient for *in vivo* rolling circle DNA replication and packaging. Eur J Biochem 149:579–584

Fluit AC, Baas PD, Jansz HS (1986) Termination and reinitiation signals of bacteriophage ϕX174 rolling circle DNA replication. Virology 154:357–368

Francke B, Ray DS (1971) Formation of the parental replicative form DNA of bacteriophage ϕX174 and initial events in its replication. J Mol Biol 61:565–586

Francke B, Ray DS (1972) *Cis*-limited action of the gene A product of bacteriophage ϕX174 and the essential bacterial site. Proc Natl Acad Sci USA 69:475–479

Fulford W, Model P (1984) Specificity of translational regulation by two DNA-binding proteins. J Mol Biol 173:211–226

Geider K, Beck E, Schaller H (1978) An RNA transcribed from DNA at the origin of phage fd single-strand to replicative form conversion. Proc Natl Acad Sci USA 75:645–649

Gilbert W, Dressler D (1968) The rolling circle model. Cold Spring Harbor Symp Quant Biol 33:473–484

Godson GN (1974) Origin and direction of ϕX174 double- and single-stranded DNA synthesis. J Mol Biol 90:127–141

Godson GN (1977) G4 DNA replication: II. Synthesis of viral progeny single-stranded DNA. J Mol Biol 117:337–351

Godson GN (1978) The other isometric phages. In: Denhardt DT, Dressler D, Ray DS (eds) The single-stranded DNA phages. Cold Spring Harbor Laboratory, Cold Spring Harbor, New York, pp 103–112

Godson GN, Barrell BG, Staden R, Fiddes JC (1978) Nucleotide sequence of bacteriophage G4 DNA. Nature 276:236–247

Gray CP, Sommer R, Polke C, Beck E, Schaller H (1978) Structure of the origin of DNA replication of bacteriophage fd. Proc Natl Acad Sci USA 76:50–53

Greenbaum JH, Marians KJ (1984) The interaction of *Escherichia coli* replication factor Y with complementary strand origins of DNA replication. Contact points revealed by DNase footprinting and protection from methylation. J Biol Chem 259:2594–2601

Grindley JN, Godson GN (1978a) Evolution of bacteriophage ϕX174: IV. Restriction enzyme cleavage map of St-1. J Virol 127:738–744

Grindley JN, Godson GN (1978b) Evolution of bacteriophage ϕX174: V. Alignment of the ϕX174 G4 and St-1 restriction enzyme cleavage maps. J Virol 27:745–753

Heidekamp F, Langeveld SA, Baas PD, Jansz HS (1980) Studies of the recognition sequence of ϕX174 gene A protein. Cleavage site of ϕX gene A protein in St-1 RFI DNA. Nucleic Acids Res 8:2009–2021

Heidekamp F, Baas PD, Van Boom JH, Veeneman GH, Zipursky SL, Jansz HS (1981) Construction and characterization of recombinant plasmid DNAs containing sequences of the origin of bacteriophage ϕX174 DNA replication. Nucleic Acids Res 9:3335–3354

Heidekamp F, Baas PD, Jansz HS (1982) Nucleotide sequences at the ϕX gene A protein cleavage site in replicative form I DNAs of bacteriophages U3, G14 and α3. J Virol 42:91–99

Hill DF, Petersen GP (1982) Nucleotide sequence of bacteriophage f1 DNA. J Virol 44:32–46

Horiuchi K (1986) Interaction between gene II protein and the DNA replication origin of bacteriophage f1. J Mol Biol 188:215–223

Horiuchi K, Zinder ND (1976) Origin and direction of synthesis of bacteriophage f1 DNA. Proc Natl Acad Sci USA 73:2341–2345

Horiuchi K, Ravetch JV, Zinder ND (1979) DNA replication of bacteriophage f1 in vivo. Cold Spring Harbor Symp Quant Biol 43:389–399

Hourcade D, Dressler D (1978) Site-specific initiation of a DNA fragment. Proc Natl Acad Sci USA 75:1652–1656

Ikeda J, Yudelevich A, Hurwitz J (1976) Isolation and characterization of the protein coded by gene A of bacteriophage ϕX174 DNA. Proc Natl Acad Sci USA 73:2669–2673

Ikeda J, Yudelevich A, Shimamoto N, Hurwitz J (1979) Role of polymeric forms of the bacteriophage ϕX174 coded gene A protein in ϕX174 RFI cleavage. J Biol Chem 254:9416–9428

Imber R, Low R, Ray D (1983) Identification of a primosome assembly site in the region of the ori 2 replication origin of the E. coli mini-F plasmid. Proc Natl Acad Sci USA 80:7132–7136

Jacob F, Brenner S, Cuzin F (1963) On the regulation of DNA replication in bacteria. Cold Spring Harbor Symp Quant Biol 28:329–347

Johnson PH, Sinsheimer RL (1974) Structure of an intermediate in the replication of bacteriophage ϕX174 deoxyribonucleic acid: the initiation site for DNA replication. J Mol Biol 83:47–61

Johnston S, Ray DS (1984) Interference between M13 and ori-M13 plasmids is mediated by a replication enhancer sequence near the viral strand origin. J Mol Biol 177:685–700

Keegstra W, Baas PD, Jansz HS (1979) Bacteriophage ϕX174 RF DNA replication in vivo. A study by electron microscopy. J Mol Biol 135:69–89

Kim MH, Ray DS (1985) Mutational mechanisms by which an inactive replication origin of bacteriophage M13 is turned on are similar to mechanisms of activation of *ras* proto-oncogenes. J Virol 53:871–878

Kim MH, Hines JC, Ray DS (1981) Viable deletions of the M13 complementary strand origin. Proc Natl Acad Sci USA 78:6784–6788

Kornberg A (1980) In: DNA replication. Freeman, San Francisco

Kornberg A (1982) In: Supplement to DNA Replication. Freeman, San Francisco

Koths K, Dressler D (1978) Analysis of the ϕX DNA replication cycle by electron microscopy. Proc Natl Acad Sci USA 75:605–609

Koths K, Dressler D (1980) The rolling circle capsid complex as an intermediate in ϕX174 DNA replication and viral assembly. J Biol Chem 255:4328–4338

Lambert PF, Waring DA, Wells RD, Reznikoff WS (1986) DNA requirements at the bacteriophage G4 origin of complementary strand DNA synthesis. J Virol 58:450–458

Lambert PF, Kawashima E, Reznikoff WS (1987) Secondary structure at the bacteriophage G4 origin of complementary-strand DNA synthesis: in vivo requirements. Gene 53:257–264

Langeveld SA, Van Mansfeld ADM, Baas PD, Jansz HS, Van Arkel GA, Weisbeek PJ (1978) Nucleotide sequence of the origin of replication in bacteriophage ϕX174 RF DNA. Nature 272:417–419

Langeveld SA, Van Mansfeld ADM, De Winter J, Weisbeek PJ (1979) Cleavage of single-stranded DNA by the A and A* proteins of bacteriophage ϕX174. Nucleic Acids Res 7:2177–2188

Langeveld SA, Van Arkel GA, Weisbeek PJ (1980) Improved method for the isolation of the A and A* proteins of bacteriophage ϕX174. FEBS Lett 114:269–272

Langeveld SA, Van Mansfeld ADM, Van der Ende A, Van de Pol JH, Van Arkel GA, Weisbeek PJ (1981) The nuclease specificity of the bacteriophage ϕX174 A* protein. Nucleic Acids Res 9:545–563

Lau PCK, Spencer JH (1985) Nucleotide sequence and genome organization of bacteriophage S13 DNA. Gene 40:273–284

Linney E, Hayashi M (1973) The two proteins of gene A of ϕX174. Nature [New Biol] 245:6–8

Marians KJ, Soeller W, Zipursky SL (1982) Maximal limits of the Escherichia coli replication factor Y effector site sequences in pBR322 DNA. J Biol Chem 257:5656–5662

Martin DM, Godson GN (1977) G4 DNA replication: I. Origin of synthesis of the viral and complementary DNA strands. J Mol Biol 117:321–335

McMacken R, Ueda K, Kornberg A (1977) Migration of Escherichia coli dnaB protein on the template DNA strand as a mechanism in initiating DNA replication. Proc Natl Acad Sci USA 74:4190–4194

Messing J, Vieira J (1982) A new pair of M13 vectors for selecting either DNA strand of double-digest restriction fragments. Gene 19:269–276

Messing J, Gronenborn B, Muller-Hill B, Hofschneider PH (1977) Filamentous coliphage M13 as a cloning vehicle: insertion of a HindIII fragment of the lac regulatory region in M13 replicative form in vitro. Proc Natl Acad Sci USA 74:3642–3646

Meyer TF, Geider K (1979a) Bacteriophage fd gene II-protein: I. Purification, involvement in RF replication, and the expression of gene II. J Biol Chem 254:12636–12641

Meyer TF, Geider K (1979b) Bacteriophage fd gene II-protein: II. Specific cleavage and relaxation of supercoiled RF from filamentous phages. J Biol Chem 254:12642–12646

Meyer TF, Geider K (1980) Replication of phage fd with purified proteins. In: Alberts B, Fox CF (eds) Mechanistic studies of DNA replication and genetic recombination. Academic, New York, pp 579–588

Meyer TF, Geider K (1982) Enzymatic synthesis of bacteriophage fd viral DNA. Nature 296:828–832

Meyer TF, Geider K, Kurz C, Schaller H (1979) Cleavage site of bacteriophage fd gene II-protein in the origin of viral strand replication. Nature 278:365–367

Model P, McGill C, Mazur B, Fulford WD (1982) The replication of bacteriophage f1: gene V protein regulates the synthessis of gene II protein. Cell 29:329–335

Moses PB, Model P (1984) A rho-dependent transcription termination signal in bacteriophage f1. J Mol Biol 172:1–22

Nomura N, Ray DS (1980) Replication of bacteriophage M13: XV. Location of the specific nick in M13 replicative form II accumulated in Escherichia coli polA ex1. J Virol 34:162–167

Nomura N, Low R, Ray DS (1982a) Identification of ColE1 sequences that direct single-strand to double-strand conversion by a ϕX type mechanism. Proc Natl Acad Sci USA 79:3153–3157

Nomura N, Low R, Ray DS (1982b) Selective cloning of ColE1 DNA initiation sequences using the cloning vector M13 Δ E101. Gene 18:239–246

Ogawa T, Arai K, Okazaki T (1983) Site selection and structure of DNA-linked RNA primers synthesized by the primosome in phage ϕX174 DNA replication in vitro. J Biol Chem 258:13353–13358

Peeters BPH, Peters RM, Schoenmakers JGG, Konings RNH (1985) Nucleotide sequence and genetic organization of the genome of the N-specific filamentous bacteriophage IKe; comparison with the genome of the F-specific filamentous phages M13, fd and f1. J Mol Biol 181:27–39

Peeters BPH, Schoenmakers JGG, Konings RNH (1986a) The gene II proteins of the filamentous phage IKe and Ff (M13, fd and f1) are not functionally interchangeable during viral strand replication. Nucleic Acids Res 14:5067–5080

Peeters BPH, Schoenmakers JGG, Konings RNH (1986b) Plasmid pKUN9, a versatile vector for the selective packaging of both DNA strands into single-stranded DNA containing phage-like particles. Gene 41:39–46

Peeters BPH, Schoenmakers JGG, Konings RNH (1987) Functional comparison of the DNA sequences involved in the replication and packaging of the viral strands of the filamentous phage IKe and Ff (M13, fd and f1). DNA 6:139–147

Ray DS, Dueber J (1975) Structure and replication of replicative forms of the ϕX-related bacteriophage G4. In: Goulian M, Hanawalt P (eds) DNA synthesis and its regulation. Benjamin, Menlo Park, California, pp 370–385

Ray DS, Cleary JM, Hines JC, Kim MH, Strathearn M, Kaguni LS, Roark M (1981) DNA initiation determinants of bacteriophage M13 and of chimeric derivatives carrying foreign replication determinants. In: Ray DS, Fox CF (eds) The initiation of DNA replication. Academic, New York, pp 169–193

Reinberg D, Zipursky SL, Weisbeek PJ, Brown DR, Hurwitz J (1983) Studies on the ϕX174 gene A protein-mediated termination of leading strand DNA synthesis. J Biol Chem 258:529–537

Roth MJ, Brown DR, Hurwitz J (1984) Analysis of bacteriophage ϕX174 gene A protein-mediated termination and reinitiation of ϕX DNA synthesis: II. Structural characterization of the covalent ϕX A protein-DNA complex. J Biol Chem 259:10556–10567

Sakai H, Godson GN (1985) Isolation and construction of mutants of the G4 minus strand origin: analysis of their in vivo activity. Biochim Biophys Acta 826:30–37

Sakai H, Komano T, Godson GN (1985) Essential structures in the complementary DNA origin of bacteriophage G4. Agric Biol Chem 49:1505–1507

Sakai H, Komano T, Godson GN (1987) Replication origin (ori$_c$) on the complementary DNA strand of Escherichia coli phage G_4: biological properties of mutants. Gene 53:265–273

Sanger F, Coulson AR, Friedman T, Air GM, Barrell BG, Brown NL, Fiddes JC, Hutchison CA III, Slocombe PM, Smith M (1978) The nucleotide sequence of bacteriophage ϕX174. J Mol Biol 125:225–246

Sanhueza S, Eisenberg S (1984) Cleavage of single-stranded DNA by the ϕX174 A* protein. The A* single-stranded DNA covalent linkage. Proc Natl Acad Sci USA 81:4285–4289

Sanhueza S, Eisenberg S (1985) Bacteriophage ϕX174 A protein cleaves single-stranded DNA and binds to it covalently through a tyrosyl-dAMP phosphodiester bond. J Virol 53:695–697

Schaller H (1979) The intergenic region and the origins for filamentous phage DNA replication. Cold Spring Harbor Symp Quant Biol 43:401–408

Schaller H, Uhlmann A, Geider K (1976) A DNA fragment from the origin of single-strand to double-strand DNA replication of bacteriophage fd. Proc Natl Acad Sci USA 73:49–53

Shlomai J, Kornberg A (1980a) An Escherichia coli replication protein that recognizes a unique sequences within a hairpin region in ϕX174 DNA. Proc Natl Acad Sci USA 77:799–803

Shlomai J, Kornberg A (1980b) A prepriming DNA replication enzyme of Escherichia coli: II. Actions of protein n': a sequence specific DNA dependent ATPase. J Biol Chem 255:6794–6798

Sims J, Benz EW Jr (1980) Initiation of DNA replication by the Escherichia coli dnaG protein: evidence that tertiary structure is involved. Proc Natl Acad Sci USA 77:900–904

Sims J, Dressler D (1978) Site-specific initiation of a DNA fragment: DNA sequence of the initiator region. Proc Natl Acad Sci USA 75:3094–3098

Sims J, Capon D, Dressler D (1979) dnaG (primase)-dependent origins of DNA replication: nucleotide sequences of the negative strand initiation sites of bacteriophages St-1, ϕK, α3. J Biol Chem 254:12615–12628

Sinsheimer RL (1959) Purification and properties of bacteriophage ϕX174. J Mol Biol 1:37–42

Smits MA, Jansen J, Konings RHN, Schoenmakers JGG (1984) Initiation and termination signals for transcription in bacteriophage M13. Nucleic Acids Res 12:4071–4081

Soeller E, Marians KJ (1982) Deletion mutants defining the Escherichia coli replication factor Y effector site sequences in pBR322 DNA. Proc Natl Acad Sci USA 79:7253–7257

Soeller E, Greenbaum J, Abarzua P, Marians KJ (1983) The interaction of Escherichia coli replication factor Y with origins of DNA replication. In: Cozarelli N (ed) UCLA symposia on molecular and cellular biology new series, vol 10. Liss, New York, pp 125–134

Soeller W, Abarzua P, Marians KJ (1984) Mutational analysis of primosome assembly sites: II. Role of secondary structure in the formation of active sites. J Biol Chem 259:14293–14300

Stayton M, Kornberg A (1983) Complexes of *Escherichia coli* primase with the replication origin of G4 phage DNA. J Biol Chem 258:13205–13212

Suggs SV, Ray DS (1977) Replication of bacteriophage M13: XI. Localization of the origin for M13 single-strand synthesis. J Mol Biol 110:147–163

Suggs SV, Ray DS (1979) Nucleotide sequence of the origin for bacteriophage M13 DNA replication. Cold Spring Harbor Symp Quant Biol 43:379–388

Tabak HF, Griffith J, Geider K, Schaller H, Kornberg A (1974) Initiation of deoxyribonucleic acid synthesis: VII. A unique location of the gap in the M13 replicative duplex synthesized *in vitro*. J Biol Chem 249:3049–3054

Van der Avoort HGAM, Van Arkel GA, Weisbeek PJ (1982) Cloned bacteriophage ϕX174 DNA sequence interferes with synthesis of the complementary strand of infecting bacteriophage ϕX174. J Virol 42:1–11

Van der Avoort HGAM, Van der Ende A, Van Arkel GA, Weisbeek PJ (1984) Incompatibility regions in the single-stranded DNA phages ϕX174, G4 and M13. J Virol 50:533–540

Van der Ende A, Teertstra R, Van der Avoort HGAM, Weisbeek PJ (1983) Initiation signals for complementary strand DNA synthesis on single-stranded plasmid DNA. Nucleic Acids Res 11:4957–4975

Van Mansfeld ADM, Langeveld SA, Weisbeek PJ, Baas PD, Van Arkel GA, Jansz HS (1979) Cleavage site of ϕX174 gene *A* protein in ϕX and G4 RFI DNA. Cold Spring Harbor Symp Quant Biol 43:331–334

Van Mansfeld ADM, Langeveld SA, Baas PD, Jansz HS, Van der Marel GA, Veeneman GH, Van Boom JH (1980) Recognition sequence of bacteriophage ϕX174 gene *A* protein: an initiator of DNA replication. Nature 288:561–566

Van Mansfeld ADM, Van Teeffelen HAAM, Zandberg J, Baas PD, Jansz HS, Veeneman GH, Van Boom JH (1982) A* protein of bacteriophage ϕX174 carries an oligonucleotide which it can transfer to the 3′-OH of a DNA chain. FEBS Lett 150:103–108

Van Mansfeld ADM, Baas PD, Jansz HS (1984a) Gene *A* protein of bacteriophage ϕX174 is a highly specific single-strand nuclease and binds via a tyrosyl residue to DNA after cleavage. Adv Exp Med Biol 179:221–230

Van Mansfeld ADM, Van Teeffelen HAAM, Baas PD, Veeneman GH, Van Boom JH, Jansz HS (1984b) The bond in the bacteriophage ϕX174 gene *A* protein-DNA complex in a tyrosyl-5′-phosphate ester. FEBS Lett 173:351–356

Van Mansfeld ADM, Van Teeffelen HAAM, Fluit AC, Baas PD, Jansz HS (1986a) Effect of SSB protein on cleavage of single-stranded DNA by ϕX gene *A* and A* protein. Nucleic Acids Res 14:1845–1861

Van Mansfeld ADM, Van Teeffelen HAAM, Baas PD, Jansz HS (1986b) Two juxtaposed tyrosyl-OH groups participate in ϕX174 gene *A* protein catalysed cleavage and ligation of DNA. Nucleic Acids Res 14:4229–4238

Van Wezenbeek PMGF, Hulsebos JJM, Schoenmakers JGG (1980) Nucleotide sequence of the filamentous bacteriophage M13 DNA genome: comparison with phage fd. Gene 11:129–148

Weisbeek PJ, Van Mansfeld ADM, Kuhlemeier C, Van Arkel GA, Langeveld SA (1981) Properties of the *A* and A* proteins of bacteriophage G4. The origin of G4 replicative-form DNA replication. Eur J Biochem 114:501–507

Westergaard O, Brutlag D, Kornberg A (1972) Initiation of deoxyribonucleic acid synthesis: IV. Incorporation of the ribonucleic primer into the phage replicative form. J Biol Chem 248:1361–1364

Wickener S, Hurwitz J (1975) Association of ϕX174 DNA-dependent ATPase activity with an *Escherichia coli* protein, replication factor Y, required for *in vitro* synthesis of ϕX174 DNA. Proc Natl Acad Sci USA 72:3342–3346

Wickner W, Brutlag D, Schekman R, Kornberg A (1972) RNA synthesis initiates *in vitro* conversion of M13 DNA to its replicative form. Proc Natl Acad Sci USA 69:965–969

Yen TSB, Webster RE (1982) Translational control of bacteriophage f1 gene II and gene X proteins by gene V protein. Cell 29:337–345

Zechel K, Bouché JP, Kornberg A (1975) Replication of phage G4. A novel and simple system for the initiation of deoxyribonucleic acid synthesis. J Biol Chem 250:4684–4689

Zinder ND, Horiuchi K (1985) Multiregulatory element of filamentous bacteriophages. Microbiol Rev 49:101–106

Zipursky SL, Marians KJ (1980) Identification of two *Escherichia coli* factor Y effector sites near the origins of replication of the plasmids Col E1 and pBR322. Proc Natl Acad Sci USA 77:6521–6524

Zipursky SL, Marians KJ (1981) *Escherichia coli* Y sites of plasmid pBR322 can function as origins of DNA replication. Proc Natl Acad Sci USA 78:6111–6115

Zipursky SL, Reinberg D, Hurwitz J (1980) *In vitro* DNA replication of recombinant plasmid DNAs containing the origin of progeny replicative form DNA synthesis of phage ϕX174. Proc Natl Acad Sci USA 77:5182–5186

Zolotukhin AS, Drygin YuF, Bogdanov AA (1984) Bacteriophage ϕX174 A* protein binds *in vitro* to the phage ϕX174 DNA by a phosphodiester bond via a tyrosine residue. Biochemistry International 9:799–806

Zuccarelli AJ, Benbow RM, Sinsheimer RL (1976) Formation of parental replicative form of bacteriophage ϕX174. J Mol Biol 106:375–402

Initiation of DNA Replication by Primer Proteins: Bacteriophage φ29 and Its Relatives

M. Salas

Centro de Biologia Molecular (CSIC-UAM), Universidad Autónoma, Campus de Cantoblanco, 28049 Madrid, Spain

Current Topics in Microbiology and Immunology, Vol. 136
© Springer-Verlag Berlin·Heidelberg 1988

1 Introduction

The fact that none of the known DNA polymerases is able to initiate DNA chains but only to elongate from a free 3'-OH group raises the problem of how replication is initiated, both at the replication origin and on Okazaki fragments. It was first shown by A. KORNBERG et al. that a general mechanism to initiate replication is through the formation of an RNA primer catalyzed by RNA polymerases or by a new class of enzymes, the primases (KORNBERG 1980). This mechanism, which can be used in the case of circular DNA molecules or linear DNAs that circularize or form concatemers, cannot be used at the ends of linear DNAs since the RNA primer is removed from the DNA chain, and there is no way of filling the gap resulting at the 5'-ends of the newly synthesized DNA chain. In some cases linear DNA molecules contain a palindromic nucleotide sequence at the 3'-end that allows the formation of a hairpin structure which provides the needed free 3'-OH group for elongation. This mechanism, first proposed by CAVALIER-SMITH (1974) for eukaryotic DNA replication, was shown to take place in several systems (KORNBERG 1980, 1982). Another mechanism to initiate replication consists in the specific nicking of one of the strands of a circular double-stranded DNA, producing a 3'-OH group available for elongation (KORNBERG 1980). In the case of the *Bacillus subtilis* phage ϕ 29, which contains a linear, double-stranded DNA of molecular weight 11.8×10^6 (SOGO et al. 1979), the initiation of replication cannot take place by any of the indicated mechanisms. In this review I will describe the existence of a protein covalently linked to the ends of ϕ 29 DNA as well as to the DNA ends of phages related to ϕ 29 and its role in the initiation of replication by a protein-priming mechanism (SALAS 1983).

2 Characterization of a Protein Covalently Linked at the 5'-Ends of ϕ 29 and Related Phage DNAs

The first indication for the existence of a protein attached to ϕ 29 DNA was the formation in vitro of circular ϕ 29 DNA molecules or concatemeres when the DNA was isolated from phage particles and the conversion of these molecules into linear ϕ 29 DNA of unit length after treatment with proteolytic enzymes (ORTIN et al. 1971). Later on, it was shown that transfection by ϕ 29 DNA was sensitive to protease treatment (HIROKAWA 1972) and that the DNA isolated from a ϕ 29 *ts* mutant in gene 3 is thermolabile for transfection (YANOFSKY et al. 1976). Indeed, it was found that a protein, characterized as the gene 3 product (SALAS et al. 1978), was covalently linked at the 5'-ends of ϕ 29 DNA (SALAS et al. 1978; ITO 1978; YEHLE 1978).

Other *B. subtilis* phages morphologically similar to ϕ 29, such as ϕ 15, ϕ 21, PZE, PZA, Nf, M2Y, B103, SF5, and GA-1, have been described and shown to contain linear, double-stranded DNAs with molecular weights $11-13 \times 10^6$.

Transfection by these DNAs is sensitive to treatment with proteolytic enzymes, suggesting the existence of a terminal protein (GEIDUSCHEK and ITO 1982; FUČIK et al. 1980). In fact, a terminal protein of a size similar to the ϕ29 p3 has been characterized in ϕ15, PZA, Nf, M2Y, B103, and GA-1 (YOSHIKAWA and ITO 1981; GUTIÉRREZ et al. 1986b). The phages have been classified into three serological classes: ϕ29, ϕ15, PZA, and PZE; Nf, M2Y, and B103; GA-1.

2.1 Nucleotide Sequence of the Gene Coding for the Terminal Protein of ϕ29 and Related Phages

The nucleotide sequence of ϕ29 gene 3, coding for the terminal protein, has been determined (ESCARMÍS and SALAS 1982; YOSHIKAWA and ITO 1982). Sequencing of the DNA region of the gene 3 mutant *sus*3(91) confirmed that the open reading frame coding for a 266-amino acid protein was, indeed, that of protein p3 (ESCARMÍS and SALAS 1982). The complete nucleotide sequence of the DNA from the ϕ29-related phage PZA, 19363 base pairs (bp) long, is now available (PAČES et al. 1985; V. PAČES, C. VLČEK, P. URBANÉK and Z. HOSTOMSKÝ, personal communication). The deduced amino acid sequence for the PZA terminal protein is very similar to that of protein p3 of ϕ29; there are only six amino acid replacements, five of which are neutral substitutions (PAČES et al. 1985). On the other hand, comparison of the chymotryptic or tryptic peptides of the terminal proteins of ϕ29, ϕ15, Nf, M2Y, B103, and GA-1 has revealed that the terminal proteins of ϕ29 and $\emptyset$15 are similar; those of Nf, M2Y, and B103 are related to each other and less related to ϕ29 than that of phage ϕ15; and the terminal protein of GA-1 is completely unrelated to that of the other phages (YOSHIKAWA and ITO 1981; GUTIÉRREZ et al., manuscript in preparation). In agreement with these results, anti p3-serum reacted to a similar extent with the terminal proteins of phages ϕ29, ϕ15, and PZA, whereas there was essentially no reaction with the terminal proteins of phages Nf, B103, and GA-1, as determined by a binding radioimmunoassay (GUTIÉRREZ et al. 1986b).

2.2 Site of the Linkage Between the ϕ29 Terminal Protein and DNA

The linkage between the ϕ29 terminal protein p3 and DNA was shown to be a phosphodiester bond between the OH-group of a serine residue in the terminal protein and dAMP, the terminal nucleotide at both 5′-ends of ϕ29 DNA (HERMOSO and SALAS 1980). To characterize which of the 18 serine residues in the protein is involved, the peptide that remained linked to the DNA after proteinase K-treatment was isolated, and the amino acid composition determined, which showed only one serine residue. This result indicates that the serine residue involved in the linkage to the DNA is located at position 232. Prediction of the secondary structure in the region around the linking site suggests that this serine residue is placed in a β-turn, probably located on the external part of the molecule, as revealed by the hydropathic values (HERMOSO et al. 1985).

3 ϕ29 DNA Replication In Vivo

3.1   Replicative Intermediates of ϕ29 DNA

The analysis by electron microscopy of the replicative intermediates isolated from ϕ29-infected *B. subtilis* indicates that replication starts at either end of the DNA, most of the time not simultaneously, and proceeds by a mechanism of strand displacement (INCIARTE et al. 1980; HARDING and ITO 1980). In addition, terminal protein was found at the ends of the parental and daughter DNA strands, in agreement with a role of protein p3 in the initiation of replication (SOGO et al. 1982).

3.2   Proteins Required for Replication In Vivo of ϕ29 DNA and the Related Phage M2

By using both *sus* and *ts* mutants of ϕ29 it was shown that genes 2, 3, 5, 6, and 17 are required for the synthesis of the viral DNA in vivo (TALAVERA et al. 1972; CARRASCOSA et al. 1976; HAGEN et al. 1976). The *ts* mutants available in genes 2, 3, 5, and 6 (TALAVERA et al. 1971; MORENO et al. 1974) were used in shift-up experiments in vivo. It was found that genes 2 and 3 are involved in initiation steps whereas genes 5 and 6 are concerned with some elongation step, although the possibility that they also play a role in initiation could not be ruled out from these experiments (MELLADO et al. 1980).

In the case of phage M2, genes G, E, and T are involved in the synthesis of the viral DNA in vivo; genes G and E correspond with the ϕ29 genes 2 and 3, respectively (MATSUMOTO et al. 1983).

Regarding the bacterial genes required for ϕ29 DNA replication, the host DNA polymerases I and III do not seem to be required since replication takes place in pol A$^-$ mutants (PEÑALVA and SALAS 1982) as well as in the presence of 6(p-hydroxyphenylazo)-uracil (TALAVERA et al. 1972), a known inhibitor of *B. subtilis* DNA polymerase III (MACKENZIE et al. 1973). Phage ϕ29 develops normally at 48° C in the *B. subtilis* replication mutants (reviewed by HENNEY and HOCH 1980) ts dnaB19, ts dnaC30, ts dnaD23, ts dnaE20, ts dnaF133, ts dnaI102, and QB1506 (dna-8132), the burst size being reduced about ten-fold in the mutants ts dnaG34 and ts dnaH151; no ϕ29 development occurs in mutant ts dnaA13 (SALAS, unpublished results). Therefore, gene *dnaA*, involved in ribonucleotide reduction, seems to be needed for ϕ29 DNA synthesis, genes *dnaG* and *dnaH* seem to be partially dispensable, and the rest of the bacterial genes tested do not seem to be required in ϕ29 DNA synthesis.

4 Initiation of ϕ 29 DNA Replication In Vitro

4.1 Formation of a Covalent Complex Between the Terminal Protein and 5′dAMP in ϕ 29 and M2

The protein-priming model for the initiation of replication was first proposed by REKOSH and coworkers (1977) for the initiation of replication of adenovirus DNA, which also has a terminal protein at the 5′-ends. It postulates that a free molecule of the terminal protein would be located at the DNA ends by protein-protein and/or protein-DNA interaction and react with the terminal deoxynucleoside triphosphate (dNTP) to form a terminal protein-deoxynucleoside monophosphate (dNMP) covalent complex that would provide the 3′-OH group needed for elongation by the DNA polymerase.

When extracts from ϕ 29-infected *B. subtilis* are incubated with [α-^{32}P] dATP in the presence of ϕ 29 DNA-protein p3 complex, a labelled protein is found with the electrophoretic mobility of p3, which is not formed when uninfected extracts are used or in the presence of anti-p3 serum (PEÑALVA and SALAS 1982; SHIH et al. 1982, 1984; WATABE et al. 1982; MATSUMOTO et al. 1983). Incubation of the ^{32}P-labelled protein with piperidine, under conditions in which the ϕ 29 DNA-protein p3 linkage is hydrolyzed, releases 5′-dAMP. In addition, the protein p3-dAMP complex formed could be elongated in vitro indicating that is is, indeed, an initiation complex (PEÑALVA and SALAS 1982). When extracts from M2-infected *B. subtilis* are used, an initiation complex between the M2 terminal protein and 5′-dAMP is also found (MATSUMOTO et al. 1983).

4.2 Protein Requirements for the Formation of the Initiation Complex in ϕ 29 and M2

Using extracts from ϕ 29 mutant-infected *B. subtilis*, it is found that those taken from *sus*3-infected *su*⁻ cells are unable to form the initiation complex, as expected from the model proposed. In addition, extracts from *sus*2-infected cells are also inactive, but the activity is restored by complementation of the *sus*2 and *sus*3 extracts, indicating that not only the gene 3 product, but also the gene 2 product, is essential for the initiation reaction in vitro (BLANCO et al. 1983; MATSUMOTO et al. 1983). Extracts from *sus*6- or *sus*17-infected *su*⁻ cells or from cells infected with a *ts*5 mutant at the restrictive temperature are active in the formation of the initiation complex in vitro, indicating that the products of genes 5, 6, and 17 are not essential for the initiation reaction (BLANCO et al. 1983; MATSUMOTO et al. 1983). All these results are in agreement with the in vivo roles of proteins p2 and p3 in initiation and of p5 and p6 in elongation (MELLADO et al. 1980).

Cistrons *G* and *E* of phage M2 seem to correspond to the ϕ 29 genes 2 and 3, respectively. In agreement with the ϕ 29 results, extracts from *sus*G- or *sus*E-infected *su*⁻ cells are inactive in the formation of the initiation complex,

but the addition of the two extracts gives rise to the initiation reaction (MATSU-
MOTO et al. 1983).

4.3 Overproduction and Purification of the ϕ29 Proteins p2 and p3

Since the amount of proteins p2 and p3 in ϕ29-infected *B. subtilis* is relatively
low, genes 2 and 3 have been cloned in an *E. coli* plasmid to overproduce
the proteins to facilitate their purification. A ϕ29 DNA fragment containing
genes 3, 4, 5, and most of 6 was cloned in plasmid pKC30 under the control
of the P_L promoter of phage λ. After heat induction of the *E. coli* cells harboring
the recombinant plasmid pKC30Al and containing a *ts* mutation in the λ repressor gene, protein p3, among others, was overproduced. The protein p3 synthesized in *E. coli* is active in the formation of the initiation complex when complemented with extracts from *sus* 3-infected cells (GARCÍA et al. 1983b). It has
been highly purified in an active form from the *E. coli* cells harboring the
gene 3-containing recombinant plasmid (PRIETO et al. 1984; WATABE et al.
1984a).
   A ϕ29 DNA fragment containing gene 2 was cloned in plasmid pBR322
and then, to place gene 2 under the control of the P_L promoter, a fragment
from this recombinant plasmid was cloned in plasmid pPLc28. A protein with
the electrophoretic mobility of p2 was overproduced in the *E. coli* cells harboring
the recombinant plasmid pLBw2 and was active in the formation of the initiation
complex when complemented with extracts from *sus*2-infected cells (BLANCO
et al. 1984). The protein p2 was highly purified in an active form from the
E. coli cells harboring pLBw2 (BLANCO and SALAS 1984).
   It has been suggested that proteins p2 and p3 probably form a complex
since they copurify when extracts from ϕ29-infected cells are used (see Sect. 4.4)
(WATABE et al. 1983; MATSUMOTO et al. 1984). To overproduce together proteins
p2 and p3, a fragment containing gene 3 was inserted in the gene 2-containing
recombinant plasmid pLBw2, so that both proteins could be synthesized in
E. coli. Extracts from cells transformed with this combined recombinant plasmid
are active in the formation of the initiation complex in the presence of ϕ29
DNA-p3 as template (L. BLANCO, J.M. LÁZARO, and M. SALAS, unpublished
results). The purification of the protein p2 and p3 complex is underway.

4.4 Activity of Purified Proteins p2 and p3 in the Formation
of the p3-dAMP Initiation Complex In Vitro

When highly purified proteins p2 and p3 are used in the presence of ϕ29 DNA-
p3 as template and [α-^{32}P]dATP, a small amount of p3-dAMP initiation complex is formed. The addition of 10–20 mM $(NH_4)_2SO_4$ greatly increases the
initiation reaction with the purified proteins (BLANCO and SALAS 1985b). The
stimulation by $(NH_4)_2SO_4$ is probably due to the formation of a complex between purified proteins p2 and p3, detected by glycerol gradient centrifugation,

as this complex appears in the presence, but not in the absence, of $(NH_4)_2SO_4$ (I. PRIETO, L. BLANCO, J.M. LÁZARO, M. SALAS, and J.M. HERMOSO, manuscript in preparation).

4.5 DNA Polymerase and 3′ to 5′ Exonuclease Activities of Protein p2

The purified protein p2, in addition to catalyzing the initiation reaction, has DNA polymerase activity when assayed on a template primer such as poly dA-(dT)$_{12-18}$ or on activated DNA (BLANCO and SALAS 1984; WATABE et al. 1984a). By in situ gel analysis the DNA polymerase activity was found in the protein p2 band (BLANCO and SALAS 1984). In agreement with these results, a partially purified DNA polymerase isolated from cells infected with a *ts* mutant in gene 2 shows greater heat lability than the one present in the wild-type preparation (WATABE and ITO 1983).

In addition, the ϕ 29 DNA polymerase p2 has a 3′ to 5′ exonuclease activity on single-stranded DNA, but not a 5′ to 3′ nuclease activity (BLANCO and SALAS 1985a; WATABE et al. 1984b). The 3′ to 5′ exonuclease, which might provide a proofreading mechanism, seems to be associated with the DNA polymerase since the two activities are heat inactivated with identical kinetics and both activities cosediment in a glycerol gradient (BLANCO and SALAS 1985a). Moreover, the initiation activity cosediments with the DNA polymerase and 3′ to 5′ exonuclease activities.

4.6 Effect of Aphidicolin and Nucleotide Analogues on the Phage ϕ 29 DNA Polymerase

Aphidicolin and the nucleotide analogues butylanilino dATP (BuAdATP) and butylphenyldGTP (BuPdGTP), known inhibitors of the eukaryotic DNA polymerase α (HUBERMAN 1981; KHAN et al. 1984), inhibit the protein-primed replication of ϕ 29 DNA-p3 in the presence of purified terminal protein p3 and ϕ 29 DNA polymerase p2. The main effect of aphidicolin occurring on the polymerization reaction is the decrease of the rate of elongation. The nucleotide analogues inhibit both the formation of the p3-dAMP initiation complex and its further elongation, the latter also due to a decrease in the elongation rate. The three drugs inhibit polymerization on activated DNA by protein p2 as well as the 3′ to 5′ exonuclease activity of the polymerase, indicating that the target of the drugs is the ϕ 29 DNA polymerase itself (BLANCO and SALAS 1986). This is in agreement with the finding that aphidicolin-resistant mutants of phage ϕ 29 map in gene 2 and have an altered DNA polymerase that reduces the sensitivity to aphidicolin (MATSUMOTO et al. 1986). The finding, that aphidicolin inhibits elongation to a larger extent than initiation whereas the nucleotide analogues inhibit equally both reactions, suggests the existence of two active sites in protein p2: one for initiation, at which dAMP is covalently linked to the OH group of serine residue 232 in the terminal protein, and the other for elongation (BLANCO and SALAS 1986).

5   Elongation of the p3-dAMP Initiation Complex with Purified Proteins p2 and p3

By using the purified terminal protein p3 and the DNA polymerase p2 with the ϕ 29 DNA-p3 complex as template, dATP, dGTP, dTTP, and ddCTP to stop elongation at nucleotides 9 and 12 from the left and right ϕ 29 DNA ends, respectively, it was shown that the p3-dAMP complex formed was specifically elongated to p3 linked to oligonucleotides 9 and 12 bases long (BLANCO and SALAS 1984; WATABE et al. 1984a). In addition, with this minimal, two-protein system, full-length ϕ 29 DNA can be synthesized in vitro (BLANCO and SALAS 1985b). The rate of elongation in this system is stimulated about three fold by the addition of 20 mM $(NH_4)_2SO_4$ (PRIETO et al., manuscript in preparation). If the only effect of $(NH_4)_2SO_4$ is to stabilize the complex formed by p2 and p3, the above results would suggest that the two proteins, the DNA polymerase and the terminal protein, may be needed together throughout elongation, in agreement with results from MATSUMOTO et al. (1984), suggesting the participation of the ϕ 29 terminal protein in elongation.

6   Isolation of Carboxy-Terminal Mutants of Protein p3 by In Vitro Mutagenesis of Gene 3. Effect of the Mutations on the In Vitro Formation of the Initiation Complex

By in vitro manipulation of gene 3-containing recombinant plasmids, two protein p3 mutants having some residues changed at the carboxyl end were obtained; these have a reduced activity on the formation of the initiation complex in vitro (MELLADO and SALAS 1982, 1983).

To study further the importance of the carboxyl end for the function of protein p3 in the initiation of ϕ 29 DNA replication, short deletion mutants at the carboxyl end of the protein have been constructed. A deletion of the last four carboxy-terminal amino acids of protein p3 reduces its activity in the formation of the initiation complex to about half the normal value. Deletions of twenty or more amino acids results in a small production of the protein and no priming activity. The addition of the last five carboxy-terminal amino acids to the latter deleted proteins gives rise to a higher production of the proteins, but no priming activity is obtained, suggesting that the region between residues 240 and 262 at the p3 carboxyl end, or part of it, might be essential for the normal functioning of the protein (ZABALLOS et al. 1986).

7 Activities of the Viral Protein p6

7.1 Overproduction and Purification of the Viral Protein p6

To study the in vitro function of the viral protein p6 in ϕ 29 DNA replication, a ϕ 29 DNA fragment containing gene 6 was cloned in plasmid pPLc28 under the control of the P_L promoter of phage λ. The overproduced protein, of molecular weight $\sim 12\,000$ by SDS-polyacrylamide gel electrophoresis, was highly purified and shown to be protein p6 by amino acid analysis and NH_2- and COOH-sequence determination. The apparent molecular weight of protein p6 is 23 600, suggesting that the native form of the protein is a dimer (PASTRANA et al. 1985).

7.2 Role of Protein p6 in the Initiation of Replication

When purified protein p6 is added to the in vitro assay with purified proteins p2 and p3 in the presence of ϕ 29 DNA-p3 complex as template and 0.25 μM (α-^{32}P] dATP, initiation complex formation is stimulated (PASTRANA et al. 1985). The stimulation is dependent on the dATP concentration since it only occurs at low dATP concentrations, suggesting that protein p6 might be affecting the K_m value for dATP. Indeed, the K_m value for dATP decreases from 6 μM in the absence of p6 to 1.2 μM in its presence (BLANCO et al. 1986). This value is further decreased to 0.4 μM in the presence of both $(NH_4)_2SO_4$ and p6 (L. BLANCO, J.M. LÁZARO, and M. SALAS, unpublished results).

7.3 Role of the Viral Protein p6 in Elongation

Protein p6 also stimulates the limited elongation reaction in the presence of ddCTP when added to purified proteins p2 and p3 and ϕ 29 DNA-p3 complex as template. The stimulation only occurs at a dATP concentration higher than 1 μM. At 1 μM or lower there was no elongation reaction, either in the absence or presence of p6. This result contrasts with the fact that the initiation reaction takes place at even 0.1 μM dATP concentration, suggesting that the ϕ 29 DNA polymerase has two active sites, one for initiation and one for elongation (BLANCO et al. 1986), as was proposed from the observed effect of aphidicolin and nucleotide analogues (see Sect. 4.6).

The activities in protein p6 that stimulate the initiation and limited elongation reactions in ϕ 29 DNA replication cosediment with the protein p6 peak in a glycerol gradient, indicating that both are present in the same protein (BLANCO et al. 1986).

When further elongation allowed to occur in the absence of ddCTP, stimulation by protein p6 to an extent similar to the case of the limited elongation reaction is obtained. On the other hand, the K_m values for dGTP in the presence or absence of p6 are alike, and the rate of elongation only increases slightly in the presence of p6. Therefore, a possible role for p6 in elongation may be to stimulate the incorporation of the first nucleotide(s) in the p3-dAMP initiation complex (BLANCO et al. 1986).

The effect of protein p6 on the elongation reaction is dependent on templates containing the ϕ 29 DNA replication origin. Also, the effect of protein p6 on elongation, unlike its effect on initiation, seems to be sequence dependent since the replication from the right end of ϕ 29 DNA-p3 is preferentially stimulated by protein p3 over that from the left end (BLANCO et al. 1986).

7.4 Binding of Protein p6 to Terminal ϕ 29 DNA Fragments

The binding of protein p6 to different DNA fragments was studied by DNase I footprinting experiments. The binding of p6 to the 269-bp *Hin*dIII L fragment, from the right end of ϕ 29 DNA, indicates the existence of protected DNA regions all along the fragment, about 22 nucleotides long, flanked by DNaseI hypersensitive sites every 24 nucleotides. A similar pattern is obtained when either of the two DNA strands in the fragment is labelled except that there is a constant displacement of 10 nucleotides in the protected regions and hypersensitive sites in one strand with respect to the other. This arrangement of protected and hypersensitive sites is also found with a 307-bp fragment from the left end of ϕ 29 DNA, except that a slightly higher concentration of protein p6 is needed to obtain it. A different pattern is seen with an internal ϕ 29 or pBR322 DNA fragment, suggesting that protein p6 recognizes specific signals at the ends of ϕ 29 DNA for its binding activity (SALAS et al. 1986). The binding of protein p6 to the ends of ϕ 29 DNA, producing some conformational change, may facilitate ϕ 29 DNA replication.

8 Overproduction and Purification of the Viral Protein p5

A ϕ 29 DNA fragment containing ORF 10 in the sequence of YOSHIKAWA and ITO (1982) and coding for a protein of molecular weight 13212 was cloned in plasmid pPLc28 under the control of the P_L promoter of phage λ. A protein of molecular weight about 13000 was labelled after heat induction. It has the same electrophoretic mobility in SDS-polyacrylamide gels as one of the proteins labelled in ϕ 29-infected minicells. The ORF 10 was shown to be the one corresponding to gene 5 by sequencing a *ts* 5 mutant (G. MARTÍN, J.M. LÁZARO, and M. SALAS, manuscript in preparation).

The protein p5 synthesized in *E. coli* cells harboring the gene 5-containing recombinant plasmid has been highly purified (MARTIN et al., manuscript in preparation). Experiments to study the activity of the purified protein in the in vitro ϕ 29 DNA replication system are underway.

9 Template Requirements for the Formation of the Initiation Complex

Using extracts from ϕ 29-infected *B. subtilis* it was found that the initiation reaction is dependent on the parental terminal protein. Proteinase K-treated ϕ 29 DNA or such DNA further treated with piperidine to remove the residual peptide remaining after proteinase K treatment are inactive. However, an intact DNA template is not needed. Protein p3-containing fragments from the left or right DNA ends are active as templates for formation of the initiation complex provided they reach a minimal size: a 26-bp fragment is active whereas a 10-bp one is essentially inactive. The activity of the latter is restored by ligation of an unspecific DNA sequence, suggesting that the low activity of the 10-bp fragment is due to its small size rather than to the lack of a specific sequence that must be recognized for the initiation reaction (GARCÍA et al. 1984).

When the purified proteins p2 and p3 are used, it can be confirmed that proteinase K-treated ϕ 29 DNA is not a template for the initiation reaction. However, piperidine-treated ϕ 29 DNA is active, although the activity is about five- to ten-fold lower than that obtained with ϕ 29 DNA-p3. The activity derives from the terminal sequences in ϕ 29 DNA since isolated piperidine-treated terminal fragments *Hind*III B and L, but not internal fragments, are active templates (GUTIÉRREZ et al. 1986b).

To study the DNA sequence requirements for the initiation reaction further, the terminal fragments *Bcl*I C and *Hind*III L, 73- and 269-bp in length from the left and right ϕ 29 DNA ends, respectively, were cloned in plasmid pKK223-3 in such a way that treatment of the recombinant plasmid pID13 with *Aha*III released the ϕ 29 DNA terminal sequences at the ends of the two fragments. The fragments, but not the circular plasmid or the plasmid linearized with *Hind*III, which place the ϕ 29 DNA terminal sequences far from the DNA ends, are active templates for the initiation and limited elongation reactions. The initiation reaction occurs at the end of the DNA fragments, as is the case with ϕ 29 DNA-p3, although the activity is about 15% of that obtained with ϕ 29 DNA-p3 (GUTIÉRREZ et al. 1986a). The fact that only terminal and not internal ϕ 29 DNA fragments are active as templates suggests the existence of specific sequences at the ϕ 29 DNA ends that allow recognition for the initiation reaction.

The above in vitro results also indicate that the parental terminal protein at the ends of ϕ 29 DNA, while not an absolute requirement, is important since, in its absence, template activity is greatly reduced. To show further the relevance of the parental terminal protein at the DNA ends, the terminal protein-DNA complex of the ϕ 29-related phages PZA, ϕ 15, Nf, B103, and GA-1 or the piperidine-treated DNAs were used as templates for the initiation reaction with the ϕ 29 proteins p2 and p3. The template activity of the terminal protein-DNA complex from phages PZA and ϕ 15, with a terminal protein very much related to that of ϕ 29, is similar to that of the ϕ 29 terminal protein-DNA complex; the activity is decreased with piperidine-treated DNA, as is the case with ϕ 29. The template activity of the terminal protein-DNA complex from phages Nf, B103, and GA-1, with a terminal protein much less related (Nf and B103) or unrelated (GA-1) to that of ϕ 29, is very low (Nf and B103) or nonexistent (GA-1). The same activity as with protein-free ϕ 29 DNA can be obtained after piperidine treatment of Nf, B103, and GA-1 DNAs (GUTIÉR-REZ et al. 1986 b). The above results suggest that the presence of a ϕ 29-related terminal protein allows the initiation reaction to occur, probably by protein-protein interaction, whereas the presence of an unrelated terminal protein prevents such an interaction, even with the DNA signals. Removal of the ϕ 29-related terminal protein decreases the template activity, indicating that protein-DNA interaction is less efficient than protein-protein interaction. Removal of the unrelated terminal protein increases the template activity by allowing the protein-DNA interaction to occur. On the other hand, preliminary results of transfection experiments indicate that, in vivo, the two parental terminal proteins are needed to produce viable phage progeny (C. ESCARMÍS, and M. SALAS, unpublished results).

9.1 Nucleotide Sequence at the Ends of ϕ 29 DNA and of Related Phages

An inverted terminal repeat 6 nucleotides long (5'-AAAGTA) exists in ϕ 29, PZA, ϕ 15, and B103 DNAs, and one 8 nucleotides long (5'-AAAGTAAG) in Nf and M2 DNAs. That of GA-1 DNA is 7 nucleotides long and different from those of the other phage DNAs (5'-AAATAGA). In addition, the sequence of the first 18 nucleotides at the left end of ϕ 29, PZA, ϕ 15, Nf, M2, and B103 DNA is identical, and there are homologies from nucleotide 19 to 50. The sequences beginning at nucleotide 51 in phages ϕ 29, PZA, and ϕ 15 are similar to each other and different to those of phages Nf, M2, and B103, which are alike. The sequence at the left end of GA-1 DNA is essentially unrelated to that of the other phages, except for the three terminal nucleotides. At the right end, the sequence from nucleotide 7 to 13 is very similar for the DNA of phages ϕ 29, PZA, ϕ 15, Nf, M2, and B103. In addition, these DNAs have an identical sequence from nucleotide 27 to 38; from nucleotide 39 onwards the sequence of Nf, M2, and B103 DNAs is completely different to that of ϕ 29, PZA, and ϕ 15 DNAs. The sequence at the right end of GA-1 DNA, like that of the left end, is unrelated to that of the other phage DNAs, except for the three terminal nucleotides and the region from nucleotide 29 to 41, which is a sequence practically matching that found from nucleotide 27 to 38 in the other phage DNAs (ESCARMÍS and SALAS 1981; YOSHIKAWA et al. 1981, 1985; GUTIÉRREZ et al. 1986 b).

9.2 Possible Role of the Inverted Terminal Repeat in the Initiation of DNA Replication of ϕ 29 and Related Phages

From the fact that protein-free DNA from phages ϕ 29, PZA, ϕ 15, Nf, B103, and GA-1 can be active as templates for the initiation of replication using the ϕ 29 proteins p2 and p3 and taking into account the arrangement at the ends of the DNAs of the different phages, a common sequence found in all the phages, except for GA-1 DNA, is the six-nucleotide long inverted terminal repeat. It is surprising that the template activity of protein-free GA-1 DNA is similar to that of the other phage DNAs. On the other hand, the common sequence is not enough to provide template specificity (GUTIÉRREZ et al. 1986a). Experiments aimed at elucidating which is the DNA sequence required for the initiation of ϕ 29 DNA replication are underway.

10 Model for the Protein-Primed Replication of ϕ29 and Related Phages

Figure 1 shows our present knowledge of the protein-primed replication of ϕ 29 and, most likely, of that of the ϕ 29-related phages. A free molecule of the terminal protein p3 interacts in vitro with the DNA polymerase p2 in the presence of NH_4^+ ions, and the p2-p3 complex is located at the ends of the ϕ 29 DNA-p3 template by protein-protein and protein-DNA interaction. In the presence of dATP, the DNA polymerase catalyzes the formation of a covalent complex between the OH-group of serine residue 232 in protein p3 and 5'dAMP.

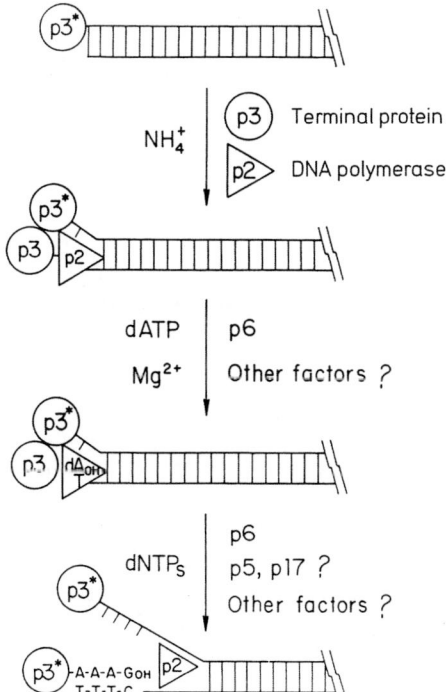

Fig. 1. Protein-primed replication of bacteriophage ϕ 29 DNA. The parental terminal protein is indicated by an *asterisk*, since it is different from free protein p3. It remains to be determined whether the terminal protein in the p3-dAMP complex resembles the parental or the free terminal protein

This reaction is stimulated in vitro by the viral protein p6 that lowers the K_m value for dATP. Whether or not other factors are involved in this initiation reaction remains to be determined. The p3-dAMP initiation complex is further elongated by DNA polymerase, which probably remains associated in a DNA replication complex. NH_4^+ ions increase about three fold the rate of elongation. The suggestion that the terminal protein also plays a role in elongation needs more investigation. The viral protein p6 stimulates the elongation reaction, probably through coupling the initiation and elongation steps by stimulating the incorporation of the first nucleotide(s) to the p3-dAMP initiation complex. The role of the viral proteins p5 and p17 in elongation in vitro as well as that of other factors, remains to be elucidated.

By using the purified protein system presently available (proteins p2, p3, and p6), unit-length $\phi 29$ DNA is obtained with $\phi 29$ DNA-p3 as template. If the parental strand is displaced before it initiates replication, formation of a panhandle structure through the six-nucleotide long inverted terminal repeat could provide a replication origin, and it might be stabilized by the terminal protein; whether or not this structure is formed remains to be determined. Another possibility is that replication at the opposite end starts before the parental strand has been completely displaced, thereby using a normal replication origin. Experiments hoping to uncover the mechanism of initiation of replication of the "displaced" parental strand are in progress.

On the other hand, although the model proposed in Fig. 1 uses a linear $\phi 29$ DNA-p3 molecule as template, an alternative possibility is the formation of circular DNA molecules by interaction of the terminal proteins at the ends of the DNA. This might explain why initiation does not occur simultaneously at the two DNA ends in vivo. In addition, the existence of the parental strand being displaced as a circle held by a protein-protein interaction might provide a mechanism for the initiation of its replication.

11   Protein-Primed Initiation of Replication: A General Mechanism

In addition to $\phi 29$ and related phages, the *Streptococcus pneumoniae* phage Cp-1 (GARCÍA et al. 1983a), the *E. coli* phage PRD1 (BAMFORD and MINDICH 1984), adenovirus (REKOSH et al. 1977), the S1 and S2 mitochondrial DNA from maize (KEMBLE and THOMPSON 1972), the linear plasmid pSLA2 from *Streptomyces* (HIROCHIKA and SAKAGUCHI 1982), and the linear plasmids pGKL1 and pGKL2 from yeast (KIKUCHI et al. 1984) have a terminal protein covalently linked at the 5-ends of the DNA. The protein-priming mechanism for initiation of replication has been studied in great detail in the case of adenovirus (STILLMAN 1983), with striking similarities to the $\phi 29$ system. Also, evidence for the in vitro formation of a covalent complex between the terminal protein and the terminal nucleotide using extracts from phage-infected bacteria has been obtained for Cp-1 and PRD1 (GARCÍA et al. 1986; BAMFORD and MINDICH 1984). It is likely that the other reported cases of DNAs with a terminal protein initiate replication also by protein priming.

In addition, several RNA genomes of animal and plant viruses have a terminal protein covalently linked at the 5'-end of the RNA (reviewed in DAUBERT and BRUENING 1984). Some evidence for the formation of a covalent complex between the terminal protein and the terminal nucleotides has been obtained for polioviral and encephalomyocarditis viral RNAs (MORROW et al. 1984; VARTAPETIAN et al. 1984). Therefore, the protein-priming mechanism is likely to be a general way to initiate replication in protein-containing nucleic acids.

Acknowledgments. The most recent work from the Madrid laboratory was carried out by A. Bernad, L. Blanco, C. Escarmis, C. Garmendia, J. Gutiérrez, J.M. Hermoso, J.M. Lázaro, G. Martín, R.P. Mellado, I. Prieto, L. Villar, and A. Zaballos. I am grateful to Dr. V. Pačes for sending me unpublished information. This research has been supported by grant 5 R01 GM27242 from the National Institutes of Health and by grants from the Comisión Asesora para el Desarrollo de la Investigación Científica y Técnica and Fondo de Investigaciones Sanitarias.

References

Bamford DH, Mindich L (1984) Characterization of the DNA-protein complex at the termini of the bacteriophage PRD1 genome. J Virol 50:309–315

Blanco L, Salas M (1984) Characterization and purification of a phage ϕ 29-encoded DNA polymerase required for the initiation of replication. Proc Natl Acad Sci USA 81:5325–5329

Blanco L, Salas M (1985a) Characterization of a 3' → 5' exonuclease activity in the phage ϕ 29-encoded DNA polymerase. Nucleic Acids Res 13:1239–1249

Blanco L, Salas M (1985b) Replication of phage ϕ 29 DNA with purified terminal protein and DNA polymerase: synthesis of full-length ϕ 29 DNA. Proc Natl Acad Sci USA 82:6404–6408

Blanco L, Salas M (1986) Effect of aphidicolin and nucleotide analogs on the phage ϕ 29 DNA polymerase. Virology, 153:179–187

Blanco L, García JA, Peñalva MA, Salas M (1983) Factors involved in the initiation of phage ϕ 29 DNA replication in vitro: requirement of the gene 2 product for the formation of the protein p3-dAMP complex. Nucleic Acids Res 11:1309–1323

Blanco L, García JA, Salas M (1984) Cloning and expression of gene 2, required for the protein-primed initiation of the *Bacillus subtilis* phage ϕ 29 DNA replication. Gene 29:33–40

Blanco L, Gutiérrez J, Lázaro JM, Bernad A, Salas M (1986) Replication of phage ϕ 29 DNA in vitro: role of the viral protein p6 in initiation and elongation. Nucleic Acids Res 14:4923–4937

Carrascosa JL, Camacho A, Moreno F, Jiménez F, Mellado RP, Viñuela E, Salas M (1976) *Bacillus subtilis* phage ϕ 29: characterization of gene products and functions. Eur J Biochem 66:229–241

Cavalier-Smith T (1974) Palindromic base sequences and replication of eukaryotic chromosome ends. Nature 250:467–470

Daubert SD, Bruening G (1984) Detection of genome-linked proteins of plants and animal viruses. Methods Virol 8:347–379

Escarmís C, Salas M (1981) Nucleotide sequence at the termini of the DNA of *Bacillus subtilis* phage ϕ 29. Proc Natl Acad Sci USA 78:1446–1450

Escarmís C, Salas M (1982) Nucleotide sequence of the early genes 3 and 4 of bacteriophage ϕ 29. Nucleic Acids Res 10:5785–5798

Fučik V, Grunow E, Grünnerová H, Hostomský Z, Zadražyl S (1980) New members of *Bacillus subtilis* phage group containing a protein link in their circular DNA. Zadražyl S, Sponar J, (eds) In DNA: recombination, interactions and repair. Pergamon, New York, pp 111–118

García E, Gómez A, Ronda C, Escarmís C, Lopez R (1983a) Pneumococcal bacteriophage Cp-1 contains a protein tightly bound to the 5' termini of its DNA. Virology 128:92–104

García JA, Pastrana R, Prieto I, Salas M (1983b) Cloning and expression in *Escherichia coli* of the gene coding for the protein linked to the ends of *Bacillus subtilis* phage ϕ 29 DNA. Gene 21:65–76

García JA, Peñalva MA, Blanco L, Salas M (1984) Template requirements for the initiation of phage ϕ 29 DNA replication in vitro. Proc Natl Acad Sci USA 81:80–84

García P, Hermoso JM, García JA, García E, López E, Salas M (1986) Formation of a covalent complex between the terminal protein of pneumococcal bacteriophage Cp-1 and 5'-dAMP. J Virol 58:31–35

Geiduschek EP, Ito J (1982) Regulatory mechanisms in the development of lytic bacteriophages in *Bacillus subtilis*. In: Dubnau DA (ed) The Molecular Biology of the Bacilli. Academic, London, 1:203–245

Gutiérrez J, García JA, Blanco L, Salas M (1986a) Cloning and template activity of the origins of replication of phage ϕ 29 DNA. Gene 43:1–11

Gutiérrez J, Vinós J, Prieto I, Méndez E, Hermoso JM, Salas M (1986b) Signals in the DNA-Terminal protein template for the initiation of phage ϕ 29 DNA replication. Virology 155:474–483

Hagen EW, Reilly BE, Tosi ME, Anderson DL (1976) Analysis of gene function of bacteriophage ϕ 29 of *Bacillus subtilis*: identification of cistrons essential for viral assembly. J Virol 19:501–517

Harding NE, Ito J (1980) DNA replication of bacteriophage ϕ 29: characterization of the intermediates and location of the termini of replication. Virology 104:323–338

Henney DJ, Hoch JA (1980) The *Bacillus subtilis* chromosome. Microbiol Rev 44:57–82

Hermoso JM, Salas M (1980) Protein p3 is linked to the DNA of phage ϕ 29 through a phosphoester bond between serine and 5'-dAMP. Proc Natl Acad Sci USA 77:6425–6428

Hermoso JM, Méndez E, Soriano F, Salas M (1985) Location of the serine residue involved in the linkage between the terminal protein and the DNA of ϕ 29. Nucleic Acids Res 13:7715–7728

Hirochika H, Sakaguchi R (1982) Analysis of linear plasmids isolated from *Streptomyces*: association of protein with the ends of the plasmid DNA. Plasmid 7:59–65

Hirokawa H (1972) Transfecting deoxyribonucleic acid of *Bacillus* bacteriophage ϕ 29. Proc Natl Acad Sci USA 69:1555–1559

Huberman JA (1981) New views of the biochemistry of eukaryotic DNA replication revealed by aphidicolin, an unusual inhibitor of DNA polymerase α. Cell 23:647–648

Inciarte MR, Salas M, Sogo JM (1980) Structure of replicating DNA molecules of *Bacillus subtilis* bacteriophage ϕ 29. J Virol 34:187–199

Ito J (1978) Bacteriophage ϕ 29 terminal protein: its association with the 5' termini of the ϕ 29 genome. J Virol 28:895–904

Kemble RJ, Thompson RD (1982) S1 and S2, the linear mitochondrial DNAs present in a male sterile line of maize, possess terminally attached proteins. Nucleic Acids Res 10:8181–8190

Khan NN, Wright GE, Dudycz LW, Brown NC (1984) Butylphenyl dGTP: a selective and potent inhibitor of mammalian DNA polymerase alpha. Nucleic Acids Res 12:3695–3706

Kikuchi Y, Hirai K, Hishinuma F (1984) The yeast linear DNA killer plasmids, pGLK1 and pGLK2, possess terminally attached proteins. Nucleic Acids Res 12:5685–5692

Kornberg A (1980) DNA replication. Freeman, San Francisco

Kornberg A (1982) DNA replication supplement. Freeman, San Francisco

Mackenzie JM, Neville MM, Wright JE, Brown NE (1973) Hydroxyphenylazopyrimidine: characterization of the active forms and their inhibitory action on a DNA polymerase from *Bacillus subtilis*. Proc Natl Acad Sci USA 70:512–516

Matsumoto K, Saito T, Hirokawa H (1983) In vitro initiation of bacteriophage ϕ 29 and M2 DNA replication: genes required for formation of a complex between the terminal protein and 5'dAMP. Mol Gen Genet 191:26–30

Matsumoto K, Saito T, Kim CI, Ando T, Hirokawa H (1984) Bacteriophage ϕ 29 DNA replication in vitro: participation of the terminal protein and the gene 2 product in elongation. Mol Gen Genet 196:381–386

Matsumoto K, Kim CI, Urano S, Ohashi H, Hirokawa H (1986) Aphidicolin-resistant mutants of bacteriophage ϕ 29: genetic evidence for altered DNA polymerase. Virology 152:32–38

Mellado RP, Salas M (1982) High level synthesis in *Escherichia coli* of the *Bacillus subtilis* phage ϕ 29 proteins p3 and p4 under the control of phage lambda P_L promoter. Nucleic Acids Res 10:5773–5784

Mellado RP, Salas M (1983) Initiation of phage ϕ 29 DNA replication by the terminal protein modified at the carboxyl end. Nucleic Acids Res 11:7397–7407

Mellado RP, Peñalva MA, Inciarte MR, Salas M (1980) The protein covalently linked to the 5' termini of the DNA of *Bacillus subtilis* phage ϕ 29 is involved in the initiation of DNA replication. Virology 104:84–96

Moreno F, Camacho A, Viñuela E, Salas M (1974) Suppressor-sensitive mutants and genetic map of *Bacillus subtilis* bacteriophage ϕ 29. Virology 62:1–16

Morrow CD, Hocko J, Navab M, Dasgupta A (1984) ATP is required for initiation of poliovirus RNA synthesis in vitro: demonstration of tyrosine-phosphate linkage between in vitro-synthesized RNA and genome-linked protein. J Virol 50:515–523

Ortín J, Viñuela E, Salas M, Vásquez C (1971) DNA-protein complex in circular DNA from phage ϕ 29. Nature New Biol 234:275–277

Pačes V, Vlček C, Urbánek P, Hostomský Z (1985) Nucleotide sequence of the major early region of *Bacillus subtilis* phage PZA, a close relative of ϕ 29. Gene 38:45–46

Pastrana R, Lázaro JM, Blanco L, García JA, Méndez E, Salas M (1985) Overproduction and purification of protein p6 of *Bacillus subtilis* phage ϕ 29: role in the initiation of DNA replication. Nucleic Acids Res 13:3083–3100

Peñalva MA, Salas M (1982) Initiation of phage ϕ 29 DNA replication in vitro: formation of a covalent complex between the terminal protein, p3, and 5'-dAMP. Proc Natl Acad Sci USA 79:5522–5526

Prieto I, Lázaro JM, García JA, Hermoso JM, Salas M (1984) Purification in a functional form of the terminal protein of *Bacillus subtilis* phage ϕ 29. Proc Natl Acad Sci USA 81:1639–1643

Rekosh DMK, Russell WC, Bellett AJD, Robinson AJ (1977) Identification of a protein linked to the ends of adenovirus DNA. Cell 11:283–295

Salas M (1983) A new mechanism for the initiation of replication of ϕ 29 and adenovirus DNA: priming by the terminal protein. Curr Top Microbiol Immunol 109:89–106

Salas M, Mellado RP, Viñuela E, Sogo JM (1978) Characterization of a protein covalently linked to the 5' termini of the DNA of *Bacillus subtilis* phage ϕ 29. J Mol Biol 119:269–291

Salas M, Prieto I, Gutiérrez J, Blanco L, Zaballos A, Lázaro JM, Martín G, Bernad A, Garmendia C, Mellado RP, Escarmís C, Hermoso JM (1987) Replication of phage ϕ 29 DNA primed by the terminal protein. In: Kelly T, McMacken R (eds) Mechanisms of DNA replication and recombination. UCLA symposia on molecular and cellular biology, new series, vol 47. Liss, New York, pp 215–225

Shih MF, Watabe K, Ito J (1982) In vitro complex formation between bacteriophage ϕ 29 terminal protein and deoxynucleotide. Biochem Biophys Res Commun 105:1031–1036

Shih MF, Watabe K, Yoshikawa H, Ito J (1984) Antibodies specific for the ϕ 29 terminal protein inhibit the initiation of DNA replication in vitro. Virology 133:56–64

Sogo JM, Inciarte MR, Corral J, Viñuela E, Salas M (1979) RNA polymerase binding sites and transcription map of the DNA of *Bacillus subtilis* phage ϕ 29. J Mol Biol 127:411–436

Sogo JM, García JA, Peñalva MA, Salas M (1982) Structure of protein-containing replicative intermediates of *Bacillus subtilis* phage ϕ 29 DNA. Virology 116:1–18

Stillman BW (1983) The replication of adenovirus DNA with purified proteins. Cell 35:7–9

Talavera A, Jiménez F, Salas M, Viñuela E (1971) Temperature-sensitive mutants of bacteriophage ϕ 29. Virology 46:586–595

Talavera A, Salas M, Viñuela E (1972) Temperature-sensitive mutants affected in DNA synthesis in phage ϕ 29 of *Bacillus subtilis*. Eur J Biochem 31:367–371

Vartapetian AB, Koonin EV, Agol VI, Bogdanov AA (1984) Encephalomyocarditis virus RNA synthesis in vitro is protein-primed. EMBO J 3:2593–2598

Watabe K, Ito J (1983) A novel DNA polymerase induced by *Bacillus subtilis* phage ϕ 29. Nucleic Acids Res 11:8333–8342

Watabe K, Shih MF, Sugino A, Ito J (1982) In vitro replication of bacteriophage ϕ 29 DNA. Proc Natl Acad Sci USA 79:5245–5248

Watabe K, Shih MF, Ito J (1983) Protein-primed initiation of phage ϕ 29 DNA replication. Proc Natl Acad Sci USA 80:4248–4252

Watabe K, Leusch M, Ito J (1984a) Replication of bacteriophage ϕ 29 DNA in vitro: the roles of terminal protein and DNA polymerase. Proc Natl Acad Sci USA 81:5374–5378

Watabe K, Leusch M, Ito J (1984b) A 3' to 5' exonuclease activity is associated with phage ϕ 29 DNA polymerase. Biochem Biophys Res Commun 123:1019–1026

Yanofsky S, Kawamura F, Ito J (1976) Thermolabile transfecting DNA from temperature-sensitive mutant of phage ϕ 29. Nature 259:60–63

Yehle CO (1978) Genome-linked protein associated with the 5′ termini of bacteriophage ϕ 29 DNA. J Virol 27:776–783

Yoshikawa H, Ito J (1981) Terminal proteins and short inverted terminal repeats of the small *Bacillus* bacteriophage genomes. Proc Natl Acad Sci USA 78:2596–2600

Yoshikawa H, Ito J (1982) Nucleotide sequence of the major early region of bacteriophage ϕ 29. Gene 17:323–335

Yoshikawa H, Friedmann T, Ito J (1981) Nucleotide sequences at the termini of ϕ 29 DNA. Proc Natl Acad Sci USA 78:1336–1340

Yoshikawa H, Garvey KJ, Ito J (1985) Nucleotide sequence analysis of DNA replication origins of the small *Bacillus* bacteriophages: evolutionary relationships. Gene 37:125–130

Zaballos A, Salas M, Mellado RP (1986) Initiation of phage ϕ 29 DNA replication by deletion mutants at the carboxyl end of the terminal protein. Gene 43:103–110

Subject Index